Mobile App Development: Crafting Innovative Digital Experiences

A Beginner's Guide to Building Your First Mobile Application

Jordan Mitchell

Table of Contents

INTRODUCTION

Mobile applications are essential tools influencing our daily lives and spur corporate innovation in an increasingly digital world. Knowing the basics of developing mobile apps is critical whether you're an aspiring developer or an entrepreneur with a game-changing idea. The comprehensive guide "Mobile App Development: Crafting Innovative Digital Experiences: A Beginner's Guide to Building Your First Mobile Application" is intended to accompany you on this thrilling adventure.

Rest assured; you don't need any prior programming skills to transform your app ideas into reality. This guide offers a straightforward, step-by-step process. We start from the very basics, exploring the various types of mobile applications and the entire development process. We then delve into the pivotal stages of planning, market research, and ideation, ensuring your app meets customer demands and stands out in the competitive app store.

Our guide leaves no stone unturned. You'll delve into the technical aspects of setting up your development environment, the essential coding skills needed to build your app, and the design principles that are key to creating intuitive and engaging user experiences. As you progress, you'll explore advanced topics like performance optimization, app security, backend integration, and rigorous testing, ensuring you're well-prepared for a successful app launch.

After reading this tutorial, you'll have the skills and self-assurance to create, release, and manage your first mobile application, transforming your creative concepts into memorable online experiences.

CHAPTER I

Get To Know Mobile App Development

The Importance of Mobile Apps

Mobile applications have completely transformed our daily lives, jobs, and interactions with the outside world. Within the last ten years, they have progressed from cutting-edge technological creations to indispensable instruments supporting numerous facets of our everyday existence. It is impossible to overestimate the significance of mobile apps in today's world since they are vital to increasing productivity, boosting economic growth, and raising the standard of living. The growing app market, which is expanding at a rate never seen before and reflects the increasing reliance on and demand for mobile solutions, further emphasizes their relevance.

Mobile apps play a wide range of roles in contemporary life, affecting almost every sector of human endeavor. They act as a link between the digital and real worlds, letting users complete various tasks remarkably quickly and easily. For example, instant messaging, video chatting, and social networking services that enable real-time worldwide connections to have transformed communication through mobile apps. With the use of apps like Facebook, Zoom, and WhatsApp, people can communicate with each other both personally and professionally, overcoming distance and promoting global connectivity.

Furthermore, mobile apps have significantly impacted business, giving organizations new methods to connect and interact with their clientele. The retail sector has completely changed due to e-commerce platforms like Amazon and eBay offering individualized and convenient buying experiences. These platforms radically alter consumer expectations and behavior by enabling consumers to explore, compare, and buy things from the comfort of their homes. Furthermore, financial transactions have been simplified by mobile banking apps, allowing users to handle their finances, pay bills, and transfer money with only a few smartphone touches.

Mobile app technology has also greatly benefited the healthcare industry. Apps for health and fitness, such as Fitbit and MyFitnessPal, assist users in keeping track of their food, exercise routine, and general health. By enabling users to consult with doctors remotely, telemedicine apps like Teladoc and Doctor on Demand have improved the efficiency of medical services and made healthcare more accessible. These apps guarantee that medical guidance and support are accessible at all times and locations and encourage healthier living.

Another industry where mobile apps have had a significant impact is education. Users of educational programs like Coursera, Khan Academy, and Duolingo have access to many information and learning materials.

These platforms allow people to learn in different ways and at their own pace while accommodating a wide range of learning styles. The ease of use and adaptability provided by educational applications have democratized education, increasing access to high-quality education for individuals from all socioeconomic backgrounds and geographic locations.

Mobile apps have revolutionized media consumption and entertainment as well. With streaming services like Netflix, Spotify, and YouTube, customers can access a limitless variety of content. With these apps, we can now get on-demand entertainment that fits our hectic schedules and has completely revolutionized how we watch and listen to music. Furthermore, interactive entertainment in the form of game apps has emerged, providing consumers with hitherto unthinkable levels of immersion.

The expanding app store reflects how commonplace and critical mobile apps are becoming in our everyday lives. There are currently over 2.7 billion smartphone users globally, and by 2023, the global mobile app market is expected to bring in over $935 billion in revenue. Numerous factors, such as the development of mobile technology, the expanding popularity of smartphones, and the growing need for digital solutions across various industries, are responsible for this spectacular expansion.

The ongoing development of mobile technology is one of the main factors propelling the expanding app market. Smartphones are now more capable and adaptable than ever, thanks to advancements in both hardware and software, which allow them to run various applications. The performance and accessibility of mobile apps have also improved with the advent of faster and more dependable mobile networks, such as 4G and 5G, giving users smooth and rapid connectivity.

The growing number of people using smartphones, especially in developing nations, has dramatically increased the user base for mobile applications. More individuals are becoming aware of mobile technology and its advantages as cell phones become more widely available and affordable. Thanks to this development, app developers now have more chances to reach underrepresented markets and meet the varied needs of a worldwide user base.

The growing need for digital solutions across various industries has also driven the expanding app market. Companies are increasingly realizing how crucial mobile apps are to growing consumer interaction, increasing operational efficiency, and maintaining competitiveness in a digital landscape that is changing quickly. Because of this, there is an increasing amount of money being invested in app development to provide unique and personalized solutions that cater to specific customer preferences and corporate needs.

In addition, the abundance of platforms and tools for app creation has reduced the entrance hurdles for would-be developers. Developers can produce high-caliber apps quickly and easily with user-friendly platforms like Xamarin, React Native, and Flutter. With the help of these technologies, app development has become more accessible, allowing both small enterprises and individuals to participate in and build the app economy.

In conclusion, there is no denying the significance of mobile applications in today's world since they are essential for a wide range of human endeavors, including business, healthcare, education, entertainment, and communication. The expanding app market highlights the revolutionary effect that mobile apps have on our daily lives. This growth is fueled by improvements in mobile technology, a rise in smartphone adoption, and an increase in the demand for digital solutions. As it develops, we anticipate even more powerful and inventive solutions from the app market, which will further improve our interactions with the digital world.

Types of Mobile Applications

As mobile technology has advanced, many kinds of mobile applications have been created, each with a unique function and meeting a range of user and technical needs. The three main types are hybrid apps, web apps, and native apps. Due to their distinct qualities, benefits, and drawbacks, each kind is best suited for particular situations. Developers and businesses must be aware of these distinctions when choosing which app to make.

Apps that are native are made to run on a particular operating system (OS), like Android or iOS. The programming languages and tools specific to each platform are used in developing these applications. While Android apps are frequently written with Java or Kotlin, iOS developers typically utilize Swift or Objective-C. Utilizing all of the features and capabilities of the target OS is one of the primary benefits of native apps. This entails utilizing platform-specific user interface (UI) features and design principles and gaining access to hardware components like the accelerometer, GPS, and camera.

Generally speaking, native applications function better than other kinds of mobile applications. They operate more effectively and have fewer performance problems since they are built especially for the device's OS. As a result, the user experience is more fluid and responsive, which is essential for high-performance applications like complicated data-driven tools or games. Furthermore, native apps can run offline, giving users continuous access to their features and content without an internet connection.

Nonetheless, there are several difficulties involved with creating native applications. The requirement to have distinct codebases for every platform is one of the main disadvantages. This implies that two separate versions, each built in a different programming language, are needed for an app that is meant to function on iOS and Android. The time and expense of development may rise dramatically as a result. Furthermore, maintaining and updating the app becomes more difficult because updates must be applied to multiple codebases. Notwithstanding these difficulties, many developers and companies favor native apps for their excellent performance and smooth OS interaction.

Web apps, on the other hand, are made to function on web browsers regardless of the operating system that powers them. JavaScript, HTML, and CSS are examples of standard web technologies used to develop these apps. Web apps do not need to be installed on the user's device because they can be viewed through a browser. Because changes are made on the server side and instantly mirrored across all users, they are accessible and straightforward to update.

Web applications have several benefits, one of which is their cross-platform interoperability. There's no need for distinct codebases for different operating systems because they can be viewed from any web browser device. This can save a lot of money on development and upkeep. Furthermore, since web apps don't require the lengthy clearance processes found in native apps, they are typically more straightforward to implement.

Web apps do, however, have certain restrictions. The primary disadvantage is their dependence on a reliable internet connection. While some web apps can operate offline through local storage and service workers, their features are more constrained overall than native apps. Online apps have different levels of access to the hardware and system resources of the device; performance is another issue. This can lead to less seamless user experiences and slower response times, especially for resource-intensive apps.

Hybrid apps aim to blend the advantages of web and native applications. These apps can be delivered through app stores and installed on devices similarly to native apps because they are essentially web apps wrapped in a native app shell. Typically, frameworks like Apache Cordova or Ionic create hybrid apps. These frameworks let developers create a single codebase using web technologies and then publish it on many devices.

The affordability of hybrid apps is one of their main benefits. Developers may target different operating systems with a single codebase, negating the need for independent development efforts. This lowers the initial cost of development and makes maintenance and upgrades easier. Furthermore, hybrid applications provide superior functionality and speed than regular web apps by leveraging plugins and APIs to access certain native functionalities.

Even with these advantages, hybrid applications have inherent drawbacks. Performance can be a significant issue, as the app relies on a web view to render content, which may not be as efficient as native rendering. While hybrid apps can access some native features, they may still fall short in fully leveraging the device's capabilities compared to fully native apps. This can result in a less optimal user experience, particularly for applications that require extensive use of hardware components or need high performance.

In conclusion, the choice between native, web, and hybrid apps depends on various factors, including the app's intended use, target audience, budget, and desired performance. Native apps offer superior performance and full access to device features but require separate development for each platform. Web apps provide broad accessibility and easier maintenance but rely on internet connectivity and have limited access to hardware. Hybrid apps strike a balance by offering cross-platform compatibility and access to some native features, but they may face performance limitations. By carefully considering these factors, developers, and businesses can choose the most appropriate type of mobile application to meet their specific needs and objectives.

The App Development Lifecycle

Developing an application from a conceptual notion to a finished, deployed application is called the app development lifecycle. This process has multiple vital steps, all necessary to guarantee that the finished product is well-designed, extensively tested, and satisfies user needs. Comprehending the lifecycle stages facilitates developers and stakeholders in managing resources, time, and expectations efficiently, from the initial idea to deployment.

The idea-generating and conceptualization stage marks the beginning of the app development lifecycle and is a collaborative effort. Here, the app's basic concept is defined, and brainstorming is done. It involves identifying a problem that the app will solve or a need it will satisfy. During this stage, stakeholders, including developers, designers, and business analysts, come together to establish the app's core goals, target audience, and important functionalities. This collaborative approach not only validates the idea by evaluating market demand, examining competitors, and finding potential possibilities and risks but also results in a distinct, workable app concept with a well-defined goal and reach.

The planning phase, after conceptualization, is centered on comprehensive project planning and requirement collection. During this phase, The team prepares a thorough project plan that details the resources, deadlines, development method, and milestones. Engaging with stakeholders to obtain specific requirements and features that the app must have is known as requirements gathering. This involves writing use cases, or user stories, that outline the key features and how users will interact with the application. The planning phase ensures that everyone in the team is aware of the goals and specifications of the project, laying the groundwork for the development phase.

The program's user interface (UI) and user experience (UX) are created during the following step: design. To visualize the layout and design of the app, wireframes, mockups, and prototypes are made at this phase. Designers collaborate closely with developers to guarantee that the app's design is both aesthetically beautiful and practical. The goal is to improve the user experience overall by designing a simple and intuitive interface. During this phase, user input is frequently requested to enhance the design and fix usability problems. The design phase is essential for creating the app's visual style and ensuring the user experience is smooth and exciting.

The design phase transitions into the development phase, which is an iterative process. In this phase, the real app code is written using the requirements and design from the other stages as a guide. Front-end and backend development are the two typical divisions of the development process. While backend development entails developing the server-side logic, databases, and APIs that support the operation of the app, front-end development concentrates on developing the user interface and putting the design components into practice. Agile approaches are frequently used by development teams, who divide the task into manageable sprints and regularly test and integrate new features. This iterative process ensures that the app develops in accordance with user needs and project goals by providing frequent feedback and updates, giving you the flexibility to adapt and respond to changing requirements.

Testing is a critical phase that follows the completion of the coding and co-occurs with development. Several testing techniques are used in this phase to find and address defects, guarantee functionality, and confirm that the application satisfies the requirements. Commonly carried out tests include unit testing, integration testing, system testing, and user acceptability testing (UAT). While integration testing ensures that many components operate together cohesively, unit testing verifies the correct functionality of each element separately. System testing assesses the application as a whole. In contrast, user acceptance testing (UAT) tests the app with actual users in a setting similar to production to confirm its functionality and performance. Delivering a high-quality app that satisfies user expectations and operates dependably requires extensive testing.

The app advances to the deployment phase following a successful testing phase. This stage entails preparing the app for public release and opening it to users. This usually entails submitting the software to app stores such as the Google Play Store and the Apple Software Store for mobile apps. Creating app store listings with thorough descriptions, screenshots, and marketing materials is a part of the submission process. To prevent rejection, strictly adhere to the rules and guidelines set forth by the app store. The app is formally released and made accessible for download after approval. In addition, deployment includes making sure the app's backend systems are prepared to handle user traffic and putting up the required infrastructure, such as servers and databases.

The app moves into the maintenance and support phase after deployment. During this phase, the app's functionality and security are enhanced through the release of updates, problem-solving, and performance monitoring. User input is not just important, it's crucial at this point. Your feedback directs future development efforts and helps identify areas that need improvement. Frequent updates guarantee that the app remains up to date and fulfills user expectations. Updating features in response to your input and industry trends, repairing bugs, and optimizing the app for new hardware and operating systems are all examples of maintenance.

The app development lifecycle is a thorough procedure that turns a basic concept into a finished, deployed program. The app's success depends on every phase, from planning and conception to design, development, testing, deployment, and maintenance. Developers may produce high-caliber apps that provide value, satisfy consumer requests, and offer a remarkable user experience by adhering to a disciplined process and concentrating on user requirements. Comprehending the principal phases of the application development lifecycle facilitates efficient resource management, mitigation of risks, and attainment of project objectives.

CHAPTER II

Ideation and Market Research

Generating App Idea

A vital first stage in the app development process is developing app ideas. It entails determining issues and creating creative fixes that a mobile application can handle. The capacity of an app to successfully address real-world issues is frequently what determines its success. Thus, creativity is crucial. Developers and entrepreneurs must use a variety of brainstorming approaches in addition to thorough problem analysis to produce workable app concepts. This section looks at problem identification, solution techniques, and efficient brainstorming techniques to produce creative app concepts.

The first step in developing app concepts is figuring out what issues prospective consumers might have. This necessitates close observation and comprehension of the target audience's requirements, problems, and day-to-day difficulties. At this point, conducting in-depth market research is crucial. Examining current apps, user feedback, and pinpointing market shortages are all part of market research. This aids in determining which problems require new solutions and which currently have ones. By conducting focus groups, interviews, and surveys with potential users, meaningful information about their needs and experiences can be gained. These exchanges highlight specific issues that can be converted into features for the app.

The next step after identifying issues is to consider possible fixes that an app might offer. This entails coming up with creative solutions to the problems that have been recognized. An efficient technique for creativity is the "problem-solution" method. This method begins by enumerating every issue found during the investigation stage. The brainstorming group comes up with as many viable answers as possible for each issue without considering their viability. Encouraging unrestricted creativity and ideas is the aim. After developing a list of potential solutions, the group can assess and hone these concepts to see which ones are the most workable and significant.

Mind mapping is another helpful brainstorming approach. Making a visual representation of concepts and their connections is called mind mapping. A key concept, such as a general issue or app category, is first positioned in the middle of a blank page to kick off the process. Related concepts, issues, and possible solutions spread out from this focal point. Every branch can be further divided into more focused concepts and subconcepts. Organizing ideas and visually examining several perspectives on an issue is made easier with the use of mind mapping. Additionally, it fosters associative thinking, which produces a larger pool of feasible app concepts by igniting one thought to inspire another.

Another proper brainstorming strategy is the SCAMPER method. Replace, Combine, Adapt, Modify, Put to another use, Eliminate, and Reverse is what SCAMPER stands for. This method promotes considering current issues and their solutions from several angles. For example, changing one component of a problem with another can result in an original solution. A novel app concept can be created by fusing two or more concepts. New ideas can also be generated by altering or adapting current solutions to improve functionality or fit new situations. By systematically applying every SCAMPER component to the identified challenges, developers can find novel app concepts that took time to emerge.

Cooperation and various viewpoints are essential for brainstorming to be productive. Including a broad team with a range of experiences, perspectives, and skill sets can significantly improve the caliber and diversity of ideas produced. Every team member contributes distinct perspectives and life experiences that can result in more thorough problem identification and innovative solutions. A culture of innovation is fostered by promoting open communication and a nonjudgmental atmosphere, guaranteeing that all ideas are considered.

Another way that can be very helpful when standard brainstorming techniques stalls is reverse brainstorming. Reverse brainstorming urges participants to consider how to create the problem or make it worse instead of how to solve it. This nontraditional method can uncover hidden details and underlying presumptions about the issue that might not have been noticed differently. After outlining these worst-case scenarios, the group can discuss potential fixes for these inflated issues, frequently resulting in creative and helpful app concepts.

One technique that focuses on getting a profound grasp of the user's perspective is called empathy mapping. This approach visually depicts what people say, think, feel, and do about a specific issue. Developers can better understand consumers' demands and problems by developing an empathy for people and considering their emotional and psychological experiences. This compassionate approach can produce valuable and significant solutions by identifying minor pain areas and generating app ideas that genuinely connect with consumers.

Ignoring how essential technology developments are for developing app concepts is impossible. Keeping up with the most recent developments and trends in technology might serve as an inspiration for new app ideas. Blockchain, augmented reality, machine learning, and artificial intelligence create new opportunities for creative apps. Innovative app concepts can arise from investigating how these technologies can solve current issues or generate novel experiences. Furthermore, examining patterns in user behavior, such as the increasing focus on wellness, sustainability, or remote work, can help direct the ideation process toward current and relevant app concepts.

Analyzing competitors' shortcomings is another excellent way to develop app ideas. By analyzing current apps on the market, developers can pinpoint flaws, restrictions, or opportunities for enhancement. Examining user reviews, ratings, and comments helps competitors discover what features current solutions appeal to and annoy their customers. Finding holes in rivals' products allows you to create an app that fills those gaps and provides a better user experience or extra features that rivals don't offer.

To summarize, developing app ideas is a thorough process that includes recognizing issues, formulating potential fixes, and utilizing various methods to encourage creativity and innovation. Developers can produce a wealth of app ideas by comprehending the target audience's needs, carrying out in-depth market research, and using brainstorming techniques like the problem-solving approach, mind mapping, SCAMPER, reverse brainstorming, and empathy mapping. Ideation is further enhanced by cooperation, various viewpoints, keeping up with technology developments, and competitive analysis. Using these endeavors, developers might generate inventive and significant application concepts that tackle practical issues and furnish significant resolutions for consumers.

Conducting Market Research

An essential phase in the app development process is conducting market research to ensure the finished product meets user and market requests. Analyzing rivals and comprehending your target market is essential to effective market research. These two elements are necessary to find opportunities, lower risks, and gain a competitive edge. This section explores the significance of market research, approaches to comprehend the target market, and strategies for rival analysis.

Successful market research starts with having a solid understanding of your target population. The particular user base the app is designed for comprises the target demographic. Developers may better cater their app's functionality, design, and marketing techniques to the needs and tastes of this audience by getting to know them well. Demographic research is the first step developers must take to understand the target audience. To do this, information on age, gender, income, education, occupation, and location must be gathered. User personas can be created Using this data—detailed profiles that reflect various target audience segments. The needs, objectives, and behaviors of potential users are highlighted in these personas, which help direct the development process.

Psychographic research goes one step further by examining the attitudes, interests, values, and lifestyles of the target audience. This study offers a more profound comprehension of the factors that drive users and impact their choices. For example, psychographic data may indicate that the app's target audience is fitness enthusiasts who highly value convenience, health, and customized experiences. Using these data, developers may create features that improve user pleasure and engagement by speaking to people personally.

Behavioral research aims to comprehend how users of a given program or technology behave. This entails examining feedback, preferences, pain spots, and usage trends from commercially available apps. Focus groups, interviews, and surveys are valuable tools for obtaining qualitative information on user behavior. Analytics tools can also monitor user interactions, giving quantitative information about how users engage with apps, what features they utilize most, and where they run into problems. This data is crucial for pinpointing problem areas and guaranteeing the app's smooth and straightforward user experience.

Analyzing competitors is the next stage in market research after the target audience has been thoroughly identified. Analyzing competitors is finding and assessing additional applications that perform comparable or identical tasks. Understanding the competitive environment, locating best practices, and seeing chances for distinctiveness are all aided by this method. Finding direct and indirect competitors is the first stage in the competitor analysis. While indirect competitors could target distinct markets or solve related requirements, direct competitors provide comparable solutions to the same target audience.

Once rivals have been identified, carefully analyzing their advantages and disadvantages is crucial. Analyzing their features, user interfaces, user experiences, price structures, and marketing plans are all part of this process. The SWOT analysis, which looks at each competitor's advantages, disadvantages, opportunities, and threats, is valuable for this kind of research. Strong brand recognition, distinctive features, and excellent user interfaces are strengths. Some such weaknesses include expensive costs, low user reviews, or restricted usefulness. Threats may come from established participants in the market or newly developed technology, while opportunities may come from holes in the market that rivals still need to fill.

User evaluations and ratings offer insightful information about the performance of rivals. By analyzing competing apps' positive and negative aspects, developers can pinpoint areas where their apps might shine. One way to create a more user-friendly interface is to address user complaints about a competitor's app's complexity. Furthermore, examining the type and frequency of updates and new features in rival applications can provide information about how actively they are being developed and maintained, as well as market strategy and resource allocation.

Another essential component of competitor analysis is benchmarking. This entails contrasting the app's suggested features and performance indicators with those of rivals. Key performance indicators (KPIs), including revenue, retention, and acquisition rates, can serve as benchmarks to help set reasonable expectations and goals. Additionally, benchmarking aids in locating industry norms and best practices that can guide the creation and promotion of the app.

Comprehending the marketing methods employed by rivals is vital for optimizing the app's market positioning. Analyzing their branding, messaging, advertising strategies, and user acquisition channels is part of this. It is possible to gain insight into how competitors connect with their audience and what appeals to users by examining their advertising campaigns, social media presence, and marketing materials. By identifying effective strategies, developers can modify and enhance their marketing tactics to draw in and keep users.

The ultimate objective of competition analysis is differentiation. Developers can distinguish their app from the competition by recognizing market gaps and chances for innovation. This could include providing better functionality, attending to unmet user requests, or using cutting-edge technologies. In addition to drawing customers, differentiation increases competitive advantage and brand loyalty.

Market research can be improved by utilizing artificial intelligence (AI), data analytics, and conventional techniques. Sophisticated analytics systems can process large datasets to find trends, patterns, and insights that manual analysis could miss. Sentiment analysis driven by AI may assess user reviews and social media comments to determine public opinion and spot new problems or possibilities. By predicting user behavior and market trends, predictive analytics can assist developers in making data-driven decisions.

They are keeping a close eye on the market and rival apps even after the release is imperative. The demand for mobile apps is dynamic, with new rivals, trends, and technology appearing regularly. Updating market research regularly guarantees that the app stays competitive and current. This entails monitoring shifts in user preferences, technological developments, and competitors' strategic maneuvers. Developers can modify their tactics, roll out new features, and maintain a significant market presence by keeping themselves updated.

To summarize, comprehensive market research is essential to an app's success. Determining the target population through behavioral, psychographic, and demographic research guarantees that the app satisfies user preferences and demands. Examining rivals offers perceptions of the competitive environment, exposing chances for uniqueness and creativity. Market research is more effective when strategies like SWOT analysis, benchmarking, and sophisticated analytics are used. By combining these efforts, developers may produce apps that stand out from the competition and provide users with outstanding value, leading to long-term success.

Defining Your App's Purpose and Goals

An essential step in the app development process is defining the objectives and purpose of your app. This phase entails developing a distinct value proposition and establishing precise goals, which, when combined, provide a road map for the app's long-term strategy, marketing, and development. These fundamental components guarantee that the application fulfills specific user requirements, distinguishes itself in a crowded market, and has quantifiable success.

Finding the main issue your app will address or the demand it will satisfy is the first step in establishing its goal. This calls for a deep comprehension of the target market's problems, which can be met through in-depth market research. Developers may determine the application's main goal by developing a sense of empathy and identifying users' issues. This section serves as the mission statement for the app, giving developers a clear path forward and a solid basis for all other decisions.

Making a unique value proposition (UVP) is the next step after determining the purpose. A UVP briefly explains the distinctive advantages and value the application provides its consumers. It answers the critical question: why should users select this app above others? Developers must emphasize the unique qualities, benefits, and experiences that make their software stand out from rivals to create an effective UVP. This could be enhanced usability, creative problem-solving, or unique functionality. The UVP should be understandable, concise, and aligned with the needs and preferences of the intended audience.

For instance, the UVP of an app that assists users in managing their money would highlight features like easy connectivity with bank accounts, real-time spending tracking, and tailored budgeting advice. An effective way to describe the UVP would be "Empowering you to take control of your finances with real-time tracking and personalized advice, all in one intuitive app." This statement speaks to the users' emotional needs by addressing their desire for simplicity and financial management and highlighting their distinctive characteristics.

The next crucial stage in determining the aims of your app is setting specific objectives. The development and marketing processes are guided by objectives, offering precise, quantifiable benchmarks to ensure the app fulfills its intended purpose. These goals should serve as success benchmarks and align with the app's overarching mission and UVP. Setting priorities for features, assigning resources, and monitoring the app's performance over time are all made more accessible with clear objectives.

Specific, Measurable, Achievable, Relevant, and Time-bound are the five SMART criteria for objectives. Well-defined goals give clarity on the precise requirements that must be met. For example, a specific goal would be to "increase the average session duration by 20% within six months" instead of a general one like "increase user engagement." Measurable goals make it possible to monitor development and evaluate results. Achievement can be measured using revenue, retention, and user acquisition rates. When considering the resources and limitations, attainable aims guarantee that goals are reasonable and reachable. Relevant goals support the app's mission and align with its purpose. Time-bound goals set deadlines, which instill urgency and allow for regular assessment.

"Acquire 10,000 new users within the first three months," "achieve a 25% monthly retention rate within six months," and "reach $50,000 in monthly recurring revenue within the first year" are a few examples of specific goals for a personal finance app. These goals offer quantifiable, measurable benchmarks that direct marketing and development efforts toward observable results.

Apart from numerical targets, qualitative targets hold significant importance. These include building a solid app community, increasing user contentment, and improving the app's usability. Surveys and user input are frequently needed to evaluate the performance of qualitative goals. One example of a qualitative goal might be to "achieve an average user satisfaction rating of 4.5 out of 5 within the first six months." These goals center on long-term loyalty and the entire user experience, both essential for long-term success.

Reaching these goals necessitates a calculated strategy that combines iterative development, ongoing user feedback, and efficient project management. Project management entails organizing, planning, and managing resources to accomplish objectives. This entails setting goals, allocating work, and monitoring advancement. Agile approaches, like Scrum or Kanban, are frequently used to manage the development process effectively, enabling flexibility and quick iterations depending on user input.

To adequately address user needs and refine the app, it is imperative to receive ongoing input from users. It's helpful to find areas for improvement and validate the features and functionalities of the app by routinely asking for feedback through surveys, user reviews, and beta testing. Developers may make data-driven decisions thanks to this feedback loop, which improves the usability and user happiness of the program.

A fundamental concept in accomplishing well-defined goals is iterative development. This method divides the development process into more manageable, smaller units called sprints. Every iteration concentrates on creating and testing particular features, enabling ongoing refinement and adjustment in response to input. Iterative development ensures that the app changes according to market trends and user needs, increasing the chances of accomplishing the goals.

To accomplish the goals of the app, marketing methods are also essential. The UVP of the app must be promoted to draw in and keep users. This entails using various platforms, including paid advertising, influencer relationships, social media, and content marketing. Marketing initiatives must align with the app's goals, focusing on the appropriate demographic and highlighting the software's unique features.

Moreover, progress evaluation and well-informed decision-making depend heavily on tracking and analyzing key performance indicators or KPIs. KPIs that shed light on the app's performance include engagement metrics, churn rate, lifetime value, and cost of acquisition of users. Regularly examining these metrics helps spot trends, expose problems, and improve tactics to ensure the app continues on track to meet its goals.

To sum up, establishing the objectives and purpose of your app is a crucial first stage in the app development process. App developers may ensure their product meets user demands and stands out in a crowded market by defining the primary problem their app will solve, developing a compelling, unique value proposition, and establishing specific, measurable goals. These components give the app's development, marketing, and long-term strategy a clear road map and direct it toward quantifiable success. Developers may accomplish their goals and provide an app that benefits users using clever marketing, iterative development, ongoing user input, and excellent project management.

CHAPTER III

Planning Your App

Project Planning and Management

An essential step in the app development process is defining the objectives and purpose of your app. This phase entails developing a distinct value proposition and establishing precise goals, which, when combined, provide a road map for the app's long-term strategy, marketing, and development. These fundamental components guarantee that the application fulfills specific user requirements and distinguishes itself successful app development project requires careful planning and management of the project. They ensure the project is executed well, stays under budget, and is completed on schedule. Developing a thorough project schedule and effectively allocating funds and resources are essential to effective project planning. These components offer a well-organized structure that directs the project team, boosts output, and reduces hazards.

One of the initial stages of project planning is developing a timeline for the project. The tasks and deadlines must be met to accomplish the project's goals are listed in a project timeline. It gives team members a visual picture of the project timeline, making it easier to comprehend due dates and the relationships between various tasks. Project managers must specify the project's goals and scope before drafting a timeframe. This is dividing the project into more manageable, smaller tasks and figuring out what has to be done in what order.

Project managers determine the tasks and then calculate the time needed for each activity. Based on the team's talents and the intricacy of the tasks at hand, this estimate ought to be reasonable. Accurate time estimates can be obtained using various methods, including expert opinion, historical data analysis, and analogical estimating. Including buffer intervals in the timetable is crucial to account for unforeseen circumstances and delays. These safety nets offer adaptability and lessen the effect of unexpected problems on the project schedule.

Project managers lay out the project timeline using a Gantt chart or other comparable visual aids after predicting the task durations. Tasks are shown on a horizontal timeline with start and finish dates, durations, and dependencies displayed on a Gantt chart. Dependencies show the relationships between functions, which must be finished before moving on to the next. The critical route, or the order of actions that establishes the project's minimal completion time, can be identified using this visual representation. Managing the critical path is essential to ensure the project stays on course and fulfills its deadlines.

Project managers frequently utilize project management software to construct and maintain deadlines in addition to the Gantt chart. Asana, Trello, Jira, and Microsoft Project are tools with sophisticated functionality for teamwork, progress monitoring, and scheduling. With the help of these technologies, the schedule may be updated and modified in real-time, guaranteeing that the project plan is up to date and correct the whole development process.

Allocating resources is yet another crucial component in managing and planning projects. It entails allocating the required resources, including workers, tools, and supplies, to every task on the project schedule. Allocating resources effectively guarantees that all jobs get the necessary inputs to be finished quickly and effectively. Determining the resources required for each activity is the first stage in the resource allocation process. This includes figuring out the essential knowledge and experience, the ideal team size, and any specialized tools or equipment.

Project managers determine which resources are needed and then evaluate their availability. This entails assessing the team's skill sets, workload, and availability. It's critical to allocate resources to balance the team's current responsibilities to prevent overworking anyone and to guarantee that they can produce high-quality work. Project managers may have to employ more staff, buy more equipment, or assign some work to outside suppliers if more resources are required.

Allocating funds is a crucial part of project planning and is strongly tied to allocating resources. An established budget guarantees that the project has the funds to finish all the activities and reach its goals. Estimating the expenses for each task and resource is necessary to create a budget. Salary, gear, software licenses, outside services, and other project-related costs may all be included in these costs. Accurate cost estimation is crucial to keep the project financially feasible and prevent budget overruns.

Project managers frequently utilize a cost breakdown structure (CBS), which classifies and itemizes all project expenses to build a thorough budget. Tracking actual spending versus the budget is made easier with the help of the CBS, which offers a comprehensive view of how money will be spent. Setting aside money for unforeseen expenses and project scope modifications is also crucial. This contingency budget enables the project to absorb unanticipated costs without endangering its long-term viability by serving as a financial cushion.

Monitoring and controlling project expenses regularly is essential to effective budget management. To keep the project on schedule, project managers compare actual spending to the allocated sums and make adjustments as necessary. This may entail dividing the money amongst projects, haggling over prices with suppliers, or adjusting the project's scope to stay within the allocated spending limit. This procedure can be streamlined, and real-time financial status updates for the project can be obtained using project management software with budget-tracking capabilities.

Planning and managing a project successfully requires teamwork and communication. All team members will be guaranteed to comprehend the project schedule, resource distribution, and financial limitations if there is clear communication. All parties involved in the project are informed about its status and any changes to the plan through regular meetings, progress reports, and updates. Collaboration platforms like Zoom, Microsoft Teams, and Slack make effective communication and seamless remote teamwork possible.

An additional crucial component of project planning is risk management. The impact of possible risks on the project budget and schedule can be reduced by identifying them and creating mitigation solutions. Projects developing apps frequently face risks related to technology, lack of resources, scope modifications, and outside variables like market or regulatory shifts. Project managers conduct risk assessments to pinpoint these hazards and create backup strategies to deal with them. By being proactive, the team can ensure that the project remains stable and that any challenges that may occur can be handled by the team with effectiveness.

To sum up, project management and planning are critical to the practical completion of app development projects. An organized framework that directs the project from beginning to end can be created by developing a thorough project timeline and adequately allocating resources and funds. Project managers may guarantee that a project stays on track by dividing it into manageable tasks, predicting durations, and using tools like Gantt charts to visualize the timeline. The project will have the resources to accomplish its goals if resources and funds are allocated effectively. Proactive risk management, frequent monitoring, and clear communication increase the chances of a project's success. App development initiatives can succeed by carefully planning, managing, and producing high-quality products that satisfy user and market demands. It is a crowded market and has quantifiable success.

Finding the main issue your app will address or the demand it will satisfy is the first step in establishing its goal. This calls for a deep comprehension of the target market's problems, which can be met through in-depth market research. Developers may determine the application's main goal by developing a sense of empathy and identifying users' issues. This goal serves as the mission statement for the app, giving developers a clear path forward and a solid basis for all other decisions.

Making a unique value proposition (UVP) is the next step after determining the purpose. A UVP briefly explains the distinctive advantages and value the application provides its consumers. It answers the critical question: why should users select this app above others? Developers must emphasize the unique qualities, benefits, and experiences that make their software stand out from rivals to create an effective UVP. This could be enhanced usability, creative problem-solving, or unique functionality. The UVP should be understandable, concise, and aligned with the needs and preferences of the intended audience.

For instance, the UVP of an app that assists users in managing their money would highlight features like easy connectivity with bank accounts, real-time spending tracking, and tailored budgeting advice. An effective way to describe the UVP would be "Empowering you to take control of your finances with real-time tracking and personalized advice, all in one intuitive app." This statement speaks to the users' emotional needs by addressing their desire for simplicity and financial management and highlighting their distinctive characteristics.

The next crucial stage in determining the aims of your app is setting specific objectives. The development and marketing processes are guided by objectives, offering precise, quantifiable benchmarks to ensure the app fulfills its intended purpose. These goals should be success benchmarks and align with the app's overarching mission and UVP. Setting priorities for features, assigning resources, and monitoring the app's performance over time are all made more accessible with clear objectives.

Specific, Measurable, Achievable, Relevant, and Time-bound are the five SMART criteria for objectives. Well-defined goals give clarity on the precise requirements that must be met. For example, a specific goal would be to "increase the average session duration by 20% within six months" instead of a general one like "increase user engagement." Measurable goals make it possible to monitor development and evaluate results. Achievement can be measured using revenue, retention, and user acquisition rates. When considering the resources and limitations, attainable aims guarantee that goals are reasonable and reachable. Relevant goals support the app's mission and align with its purpose. Time-bound goals set deadlines, which instill urgency and allow for regular assessment.

"Acquire 10,000 new users within the first three months," "achieve a 25% monthly retention rate within six months," and "reach $50,000 in monthly recurring revenue within the first year" are a few examples of specific goals for a personal finance app. These goals offer quantifiable, measurable benchmarks that direct marketing and development efforts toward observable results.

Apart from numerical targets, qualitative targets hold significant importance. These include building a solid app community, increasing user contentment, and improving the app's usability. Surveys and user input are frequently needed to evaluate the performance of qualitative goals. One example of a qualitative goal might be to "achieve an average user satisfaction rating of 4.5 out of 5 within the first six months." These goals center on long-term loyalty and the entire user experience, both essential for long-term success.

Reaching these goals necessitates a calculated strategy that combines iterative development, ongoing user feedback, and efficient project management. Project management entails organizing, planning, and managing resources to accomplish objectives. This entails setting goals, allocating work, and monitoring advancement. Agile approaches, like Scrum or Kanban, are frequently used to manage the development process effectively, enabling flexibility and quick iterations depending on user input.

To adequately address user needs and refine the app, it is imperative to receive ongoing input from users. It's helpful to find areas for improvement and validate the features and functionalities of the app by routinely asking for feedback through surveys, user reviews, and beta testing. Developers may make data-driven decisions thanks to this feedback loop, which improves the usability and user happiness of the program.

A fundamental concept in accomplishing well-defined goals is iterative development. This method divides the development process into more manageable, smaller units called sprints. Every iteration concentrates on creating and testing particular features, enabling ongoing refinement and adjustment in response to input. Iterative development ensures the app changes according to market trends and user needs, increasing the chances of accomplishing the goals.

To accomplish the goals of the app, marketing methods are also essential. The UVP of the app must be promoted to draw in and keep users. This entails using various platforms, including paid advertising, influencer relationships, social media, and content marketing. Marketing initiatives must align with the app's goals, focusing on the appropriate demographic and highlighting the software's unique features.

Moreover, progress evaluation and well-informed decision-making depend heavily on tracking and analyzing key performance indicators or KPIs. KPIs that shed light on the app's performance include engagement metrics, churn rate, lifetime value, and cost of acquisition of users. Regularly examining these metrics helps spot trends, expose problems, and improve tactics to ensure the app continues to meet its goals.

To sum up, establishing the objectives and purpose of your app is a crucial first stage in the app development process. App developers may ensure their product meets user demands and stands out in a crowded market by defining the primary problem their app will solve, developing a compelling, unique value proposition, and establishing specific, measurable goals. These components give the app's development, marketing, and long-term strategy a clear road map and direct it toward quantifiable success. Developers may accomplish their goals and provide an app that benefits users using clever marketing, iterative development, ongoing user input, and excellent project management.

Wireframing and Prototyping

Prototyping and wireframing are crucial phases in the app development process that aid in converting concepts into realized designs and interactive encounters. Before moving on to full-scale development, these stages allow developers and designers to see how the app will look and feel regarding its functionality and user experience. To create mobile applications that are both efficient and easy to use, it is essential to grasp the fundamentals of wireframing as well as the tools and processes of prototyping.

The first stage of the design process, wireframing, is when the app's basic framework is developed. It entails creating a low-fidelity prototype or blueprint of the application's UI, emphasizing structure, organization, and element placement. Wireframing is primarily used to map out an application's functionality and user flow without becoming bogged down with design elements like fonts, colors, and graphics. This streamlined method facilitates rapid ideation iterations and structural app refinement based on user requirements and business goals.

Knowing the information architecture and user flow is the first step toward learning the fundamentals of wireframing. Whereas information architecture focuses on arranging and identifying content to make it easy to traverse, user flow describes the route users follow to accomplish tasks within an application. The app's well-planned user flow and unambiguous information architecture guarantee that users can interact with it intuitively and effectively to achieve their objectives. Wireframes serve as a visual aid for these components, enabling designers to organize the navigation and structure of the application efficiently.

Drawing out the interface of an application on paper or using digital tools is known as creating wireframes. To depict the app's structure, essential elements, including headers, footers, buttons, menus, and content sections, are arranged in their proper locations. Prioritizing clarity and simplicity at this point is crucial, with usefulness taking precedence over looks. Various techniques can be used to produce wireframes, including hand-drawn sketches and digital tools like Balsamiq, Sketch, and Figma. These tools make generating and editing wireframes simpler by providing pre-designed pieces and drag-and-drop capability.

One of wireframing's primary benefits is that it makes early feedback and collaboration easier. Presenting wireframes to stakeholders allows designers to get critical feedback on the functionality and layout of the app. This cooperative strategy guarantees that the app satisfies user needs and aligns with corporate objectives. Furthermore, wireframes facilitate clear comprehension of the app's structure and functionality among designers, developers, and other team members and act as a communication tool.

Prototyping comes next when the wireframes are complete. Creating high-fidelity, interactive prototypes of the app that mimic the user experience is known as prototyping. By including interactive features, design components, and transitions, prototypes give wireframes life and enable users to engage with the app much like they would with the finished product. Prototyping is mainly used to evaluate the app's functionality, design, and usability before devoting time and resources to development.

Prototyping can be done with various tools and methods, each with unique features and functionalities. Prototyping software such as Adobe XD, InVision, Figma, and Axure RP are widely used. These tools offer a variety of features, ranging from basic click-through prototypes to intricate, fully interactive models. For example, Adobe XD enables designers to produce interactive prototypes that realistically depict the app's user experience using transitions, animations, and micro-interactions. Similar functionalities are available in InVision, along with real-time feedback and iteration capabilities through collaborative tools.

Adding design components like colors, fonts, photos, and icons to the wireframes in the selected tool is the first step in the prototyping process. By converting the wireframes into high-fidelity mockups, this stage gives the application a professional appearance. Interactive components like buttons, links, and form fields are incorporated to mimic user activities. Transitions and animations are also included for a seamless and captivating user experience. For example, designers can use swipe movements, hover effects, and dynamic screen transitions to imitate in-person interactions with the app.

To get input and find usability problems, user testing is another step in the prototyping process. Presenting the prototype to a group of intended users and seeing how they engage with the app is known as user testing. This procedure aids in locating trouble spots, problems with navigation, and potential trouble spots for users. User testing feedback is crucial for improving the functionality and design of the app, resulting in a seamless and simple user experience for end users. During prototyping, methods such as A/B testing can also be employed to compare several design versions and identify the best-performing one.

Iterative development is vital to the prototype process. Iterative development continuously enhances the prototype in response to user feedback and test findings. This method lowers the possibility of expensive adjustments during development by ensuring that the app develops in line with user needs and market trends. Version control is frequently supported by prototyping tools, which let designers keep track of modifications, return to earlier iterations, and try various design approaches.

Early in the design process, low-fidelity prototypes—also called paper prototypes—can be utilized in addition to high-fidelity prototypes. Rough designs of the app's interface are made on paper and used to mimic user interactions in paper prototyping. This method is quick economical and enables testing and iteration quickly. Paper prototypes help generate ideas and get preliminary feedback, even though they lack the realism and interaction of digital prototypes.

Additionally, prototyping is essential for conveying design intent to developers. A well-designed prototype ensures that developers know what's expected of them and reduces ambiguity by giving a clear and detailed picture of the app's functionality and design. The development process is streamlined, rework is reduced, and the overall quality of the finished product is improved when designers and developers work together harmoniously.

To sum up, wireframing and prototyping are essential phases in the creation of apps that aid in transforming concepts into designs that are both practical and user-friendly. Creating a skeletal framework for the application's interface while concentrating on layout, organization, and user flow is known as wireframing. It offers a precise road map that directs the design and development process. Conversely, prototyping entails building high-fidelity, interactive application versions that mimic the user interface. Prototypes enable early feedback and iterative development by validating the app's functionality, design, and usability. Wireframing and prototyping are supported by various tools and methods, from hand-drawn sketches to sophisticated digital tools, allowing designers to produce functional and captivating mobile applications. By going through these phases, developers may ensure that their applications satisfy users' needs, provide a flawless user experience, and succeed in a cutthroat market.

Creating User Stories and Flowcharts

Creating user stories and flowcharts is essential for mapping out user journeys and defining user interactions. This process is crucial in developing a user-centered design, ensuring that the final product meets the needs and expectations of its users. This section will delve into the importance of user stories and flowcharts, their creation process, and how they contribute to a seamless user journey by defining user interactions effectively.

User stories are short, simple descriptions of a feature told from the end user's perspective. They are written in everyday language and are designed to capture what the user wants to achieve using the product. User stories are a critical component of agile development methodologies, helping teams to understand the user's needs and prioritize development tasks accordingly. By focusing on the user's perspective, user stories ensure that the development process remains user-centric, ultimately leading to a product that delivers real value to its users.

Creating user stories begins with identifying the different types of users, or personas, who will interact with the product. Personas are fictional characters representing various user types within a targeted demographic. These personas are developed based on user research, including interviews, surveys, and observations. By understanding these personas' needs, goals, and pain points, teams can create user stories that accurately reflect the user's journey.

A typical user story follows a simple template: "As a [user type], I want to [action] so that [outcome]." This format ensures that the user story is concise and focused on the user's needs. For example, "As a new user, I want to sign up for an account to access exclusive content." This user story highlights the user's goal (signing up for an account) and the desired outcome (accessing exclusive content). By breaking down the user's journey into a series of user stories, teams can map out the entire user experience, identifying fundamental interactions and touchpoints.

Flowcharts, however, are visual representations of the steps involved in a process. In the context of user journeys, flowcharts help to illustrate the sequence of interactions a user will have with the product. They provide a clear and detailed view of the user's path, highlighting decision points, actions, and outcomes. Flowcharts help identify potential bottlenecks or areas where the user experience can be improved.

Creating a flowchart starts with identifying the critical stages of the user journey. These stages align with the major milestones or touchpoints in user stories. For example, the flowchart for a new user signing up for an account might include stages such as accessing the sign-up page, entering personal information, verifying email, and completing the registration process. Each stage is then broken down into individual steps, with arrows indicating the actions and decisions flow.

Creating flowcharts involves collaboration among team members, including designers, developers, and product managers. By working together, the team can ensure that the flowchart accurately reflects the user's journey and identifies potential issues or challenges. This collaborative approach also helps to align the team's understanding of the user experience, fostering a shared vision of the final product.

Flowcharts and user stories complement each other, providing a comprehensive view of the user journey. While user stories capture the user's goals and motivations, flowcharts visualize the steps required to achieve those goals. Together, they provide a detailed user experience map, helping teams identify and address potential pain points.

One of the key benefits of using user stories and flowcharts is that they facilitate effective communication among team members and stakeholders. By presenting the user journey clearly and concisely clearly and concisely, these tools help ensure that everyone involved in the project has a shared understanding of the user's needs and expectations. This shared understanding is crucial for making informed decisions throughout development, from prioritizing features to designing the user interface.

Another advantage of user stories and flowcharts is that they support iterative development. In an agile environment, teams work in short development cycles, continually refining and improving the product based on user feedback. User stories and flowcharts provide a flexible framework for this iterative process, allowing teams to adapt to changing requirements and priorities quickly. By regularly revisiting and updating these tools, teams can ensure that the product remains aligned with the user's needs and expectations.

User stories and flowcharts also play a critical role in usability testing. These tools help identify essential testing and evaluation areas by mapping the user journey. Usability testing involves observing real users as they interact with the product and identifying any issues or challenges they encounter. By comparing the observed user behavior with the expected user journey outlined in the user stories and flowcharts, teams can pinpoint specific areas for improvement. This iterative testing and refinement process helps to ensure that the final product delivers a seamless and intuitive user experience.

In conclusion, creating user stories and flowcharts is vital in mapping out user journeys and defining user interactions. By capturing the user's goals and visualizing the steps required to achieve them, these tools provide a comprehensive view of the user experience. They facilitate effective communication, support iterative development, and play a crucial role in usability testing. Ultimately, user stories and flowcharts help to ensure that the final product meets the needs and expectations of its users, delivering real value and a seamless user experience. Through careful planning and continuous refinement, teams can create products that meet and exceed user expectations, fostering satisfaction and loyalty.

CHAPTER IV

Designing User Experience (UX)

Principles of UX Design

The basis for developing digital goods that are useful, easy to use, entertaining, and available to a diverse user base is laid by UX design principles. These concepts are essential to ensure that our products are inclusive and user-centered. They involve comprehending user demands and designing for usability and accessibility. In this section, we will go into great detail about these concepts, looking at how the user's needs influence the design process and how prioritizing usability, and accessibility improves the user experience.

Effective UX design is built on a foundation of understanding user needs. This entails developing a thorough understanding of the objectives, driving forces, actions, and problems experienced by the users who will engage with the product. To arrive at this insight, designers use a range of research techniques, including focus groups, surveys, user interviews, and observational studies. These techniques yield quantitative and qualitative data that designers may utilize to build comprehensive user personas that accurately reflect user demographics.

Designers can better understand their users using user personas, made-up characters based on actual data. Designers can make well-informed judgments about the features, layout, and functionality of their products by considering these personas' demands and preferences. A persona could be a busy professional who wants to get important information quickly or a user who is blind and needs assistive technologies to use the product. By being aware of these many needs, designers can ensure that the design process stays user-centric and concentrates on developing solutions that satisfy users' needs.

Designing for usability comes next after user needs are determined. The term "usability" describes how simple and effective a product is in helping users accomplish their objectives. A well-designed product has a simple, unambiguous structure and easy navigation. When designing for usability, designers adhere to several essential rules.

For starters, simplicity is essential. The information and options should be presented briefly with no needless complexity. This calls for simple language, uncomplicated directions, and a straightforward visual design. By focusing on essential tasks and reducing distractions, users can reduce cognitive load and achieve their goals more quickly.

Second, design consistency is crucial. Users are given a sense of familiarity when colors, typefaces, and interaction patterns are used consistently throughout the product. Familiar components help users feel more confident and efficient since they can predict how the product will operate. For instance, putting navigation menus in the same spot on every page makes it easier for consumers to find what they're looking for.

Another crucial component of usability is feedback. Users must know if their actions were practical or if there were mistakes. Giving customers fast, unambiguous feedback —such as error messages with instructions on fixing problems or confirmation messages for successful actions —makes them feel more in control of their interactions with the product.

For usefulness, error recovery and prevention are equally essential. It is desirable to design interfaces that stop errors before they happen. Features that direct users toward the right behaviors, such as auto-correction, suggestions, and limitations, can help achieve this. But if mistakes are made, giving users simple means of recovery —like the ability to cancel actions or see unambiguous error messages—guarantees that they can rapidly resume their tasks.

One essential tenet of UX design is accessibility, in addition to usability. Products that are accessible are guaranteed to be useable by individuals with a variety of skills and limitations. Designing for accessibility is making goods usable by all users, including those with cognitive, motor, visual, or aural disabilities.

Providing alternate text for photos is one of the main things to remember when designing for accessibility. Users who are blind or visually challenged can read this text and understand the meaning of the photos. Similarly, for users with motor impairments who cannot operate a mouse, it is imperative to ensure that all interactive elements—such as buttons and links—can be accessed via keyboard navigation.

Yet another crucial component of accessibility is color contrast. The text should be sufficiently contrasted with its background to allow individuals with color blindness or impaired vision to read it. Color contrast ratios can be verified by designers using tools to make sure they adhere to accessibility guidelines.

Moreover, for audio and video information, designers must supply transcripts and captions. This enables people who are hard of hearing or deaf to view multimedia. Furthermore, creating unambiguous and standardized navigation architecture facilitates the understanding and navigation of the product by people with cognitive impairments.

Ensuring that a product is compatible with assistive devices, including screen readers and speech recognition software, is another aspect of accessibility. This calls for the use of semantic HTML elements, which give valuable details about the composition and content of the page, facilitating the interpretation and user presentation of the content by assistive technology.

By integrating usability and accessibility into the design process, designers produce inclusive and valuable products. This method improves the experience for all users and those with disabilities. For instance, high-contrast font can make text easier to see for individuals with different eyesight levels, and captions on movies can be helpful in noisy surroundings.

In conclusion, developing user-centered goods requires adhering to UX design principles, which include comprehending user needs, designing for usability, and guaranteeing accessibility. Through comprehensive user research and the creation of well-defined personas, designers acquire a valuable understanding of their clientele's varied requirements and inclinations. The product will be easy to use and efficient if usability characteristics like consistency, simplicity, feedback, and error prevention are prioritized. An inclusive and useable product by individuals with various abilities and disabilities is another benefit of designing for accessibility. Following these guidelines will help designers make products that offer a smooth, entertaining, and inclusive user experience, which will eventually increase customer happiness and help the product succeed in the marketplace.

Creating User Personas

The creation of user personas is a fundamental approach in design and development, offering a powerful tool for understanding and empathizing with a product or service's target audience. These personas, also known as buyer or customer personas, are not just fictional characters, but strategic tools that reflect the traits, requirements, objectives, and actions of real users. By using personas to create comprehensive user profiles, teams can align their design and development efforts with user-centric concepts, leading to more successful and user-friendly products.

Collecting and evaluating information about the intended audience is a necessary step in creating comprehensive user personas. Usually, this approach begins with a thorough investigation involving market analysis, user testing, interviews, surveys, and interviews. The objective is to gather quantitative and qualitative data that sheds light on the characteristics, desires, driving forces, problems, and habits of the users who will engage with the product. By collecting varied viewpoints and feedback from prospective users, groups can develop thorough personas that precisely reflect user requirements and experiences.

Finding recurring themes and patterns in the data to create unique user personas comes after the research process is over. Rather than presumptions or prejudices, each persona needs to be grounded in actual information and ideas. Personas, for instance, could contain details like a person's age, gender, profession, degree of schooling, and household income. Psychographic information like likes, pastimes, values, attitudes, and lifestyle preferences may also be included. Furthermore, to create realistic and relevant personas, behavioral insights pertaining to technology usage, purchasing habits, obstacles, and aspirations associated with the product or service are essential.

Creating user personas needs to be well thought out and validated to ensure correctness and relevance. It frequently combines and condenses complex data into precise, short profiles encapsulating various customer segments. Every persona ought to be unique and unforgettable, embodying a particular archetype of the intended audience and covering a range of traits and requirements that correspond with actual consumers.

After being developed, user personas are valuable tools that help direct design choices through the product development lifecycle. Teams may better understand consumers' wants and preferences using user personas, which offer a human-centered viewpoint. This helps teams make design decisions that put the demands of usability, functionality, and the overall user experience first. Personas, for example, might affect choices about visual design, content strategy, navigation, feature prioritization, and interface design.

User personas play a crucial role in fostering innovation and success by facilitating effective communication and collaboration among cross-functional teams. By using personas during brainstorming sessions, meetings, and decision-making processes, teams can synchronize their efforts and maintain focus on serving the demands of their target audience. This shared understanding of the users, their values, and likely interactions with the product, helps develop a user-centric strategy that fosters innovation and success.

Personas can help determine and rank use cases and user requirements according to various user segments' distinct requirements and objectives. For example, personas may show individual preferences or pain spots that demand specific solutions or features within the product. Teams can develop more relevant and personalized experiences that connect with users and set the product apart in the market by catering to the specific demands of each persona.

The process of creating user personas is not a one-time task, but an ongoing, iterative process. To ensure their validity and usefulness over time, personas need to be regularly updated with fresh information and insights. As products change and user preferences shift, personas should be refined to match current trends and behaviors. This can be done through persistent user involvement, feedback loops, usability testing, and analytics, which provide valuable insights for persona validation and improvement.

In conclusion, designing user personas is a planned and iterative process that entails developing complete user profiles based on research and data analysis. User personas help teams understand and sympathize with the target audience, influencing design decisions that emphasize user needs and preferences. By representing the many traits, actions, and objectives of actual users, personas help teams build more prevalent and easily navigable products that appeal to their target market. Personas are essential for human-centered design because they promote teamwork, help with decision-making, and stimulate creativity. This improves the user experience and helps products and services succeed.

Testing UX Designs

A crucial step in developing user-centered products is testing UX designs using usability tests and iterating based on user feedback. This iterative loop eventually produces a more user-friendly and efficient interface by ensuring the design satisfies user wants and expectations. This section will discuss the value of usability testing, how these tests are carried out, and how iterating a product based on user feedback may significantly improve its quality.

A crucial step in the UX design process is usability testing. It entails testing a product on actual users to assess it. With this technique, designers may better comprehend user behavior with the product, spot usability problems, and collect quantitative and qualitative data to guide their decisions. Usability testing's main objective is to ensure the product is user-friendly, effective, and fulfilling for the intended market. Designers can learn a great deal about the product's strengths and weaknesses by watching users accomplish tasks. These insights can subsequently be addressed through iterative design enhancements.

Setting precise goals is the first step in the usability testing process. These goals should align with the product's overarching objectives and concentrate on certain user experience elements that require assessment. A usability evaluation could focus on things like how simple it is to navigate, how well the user interface works, or how clear the instructions are. Making a test plan comes next when the objectives are decided upon. This plan describes the objectives of the test, the tasks that users will complete, the success metrics, and the procedure for gathering and evaluating data.

Appropriate participant selection is essential to successful usability testing. Since their opinions will have the most bearing on the design, participants should represent the product's intended market. A wide range of viewpoints can be obtained by enlisting a diverse user base, which can aid in identifying problems that may impact various user groups. Following their selection, participants are usually required to finish tasks under the testing team's observation. The assignments must mirror actual situations consumers can encounter when utilizing the software.

Observers watch how consumers interact with the product during the usability test and pay special attention to any problems or confusion. Establishing a welcoming and accepting environment for participants is critical. Hence, they feel free to express their ideas and think aloud while using the product. This verbal input offers a thorough grasp of the user's experience when paired with observational data. You may employ other tools like eye-tracking sensors or screen recording software to get more precise statistics.

Finding trends and reoccurring problems that users run into is part of the analysis process for the data gathered during usability testing. This study aids in identifying particular product areas that require improvement. Task completion rate, time spent on task, mistake rate, and user satisfaction scores are typical metrics used in usability testing. By assessing these metrics, designers can put a number on the product's usability and rank bugs according to how they affect the user experience.

After completing the usability test and data analysis, the design will be iterated upon in response to user feedback. This iterative approach is essential to make improvements to the product and ensure it fulfills user needs. Designers begin by resolving the most critical problems found during testing. This could entail improving the instructions' clarity, streamlining processes, or redesigning some components. We ensure that all design changes align with user feedback by using the insights gathered from usability testing to guide each modification.

Following the required design modifications, the updated product is put through another round of usability testing. Testing and improvement iteratively continue until the product reaches a high degree of usability. Every iteration offers a chance to get additional user input and make little adjustments that add to a gradually better overall user experience. This procedure not only assists in locating and resolving usability problems but also cultivates a constant improvement mindset within the design team.

Iterating in response to user feedback puts the needs of the user front and center in the development process, which makes it a powerful approach to UX design. Designers may make the product more intuitive and enjoyable by continuously improving it based on user interactions. This user-centered approach fosters user loyalty and trust because users believe their needs are being met and their feedback is respected. Furthermore, identifying usability problems early in the development process lowers the likelihood of costly redesigns later.

This iterative procedure can increase usability and open up new avenues for creativity. Designers may see unmet needs or develop fresh ideas to improve the user experience while watching consumers engage with the product. These realizations may spur the creation of fresh features or the improvement of existing ones, advancing the product's development and preserving its competitiveness.

Iterative design and usability testing are not restricted to a product's initial development stage. These are continuous procedures that last the whole life of the product. As user demands and preferences grow, continuous usability testing ensures that the product remains relevant and practical. By maintaining a high usability level throughout time, designers can adjust to evolving trends and shifting user behaviors thanks to this constant feedback loop.

In conclusion, a crucial step in developing user-centered products is putting UX concepts through usability testing and revising in response to user feedback. Designers can discover opportunities for improvement and obtain critical insights into the product's usability by incorporating actual users in the review process. A more user-friendly and fulfilling experience is produced by the iterative cycle of testing and improvement, which ensures the product satisfies its consumers' needs and expectations. This method produces a product that truly adds value for its users by encouraging innovation and ongoing improvement in addition to improving usability. Usability testing and iterative design can significantly enhance a product's performance by guaranteeing it stays user-friendly and efficient throughout its lifecycle through meticulous planning, execution, and analysis.

CHAPTER V

Designing User Interface (UI)

Principles of UI Design

Creating aesthetically pleasing and functional interfaces that improve user experience requires a firm grasp of UI (User Interface) design principles. Visual hierarchy, consistency, color theory, and typography are essential aspects of these principles. These components aid in directing users' attention, guaranteeing a unified appearance and feel, and enhancing the interface's usability and aesthetic appeal. We will go into great detail about these ideas in this section and discuss how they improve UI design.

One of the core ideas of UI design is visual hierarchy. It describes how components are arranged and presented to highlight their significance. Designers can direct users' attention toward the most crucial pieces by adjusting size, color, contrast, alignment, and spacing. For instance, A bolder and larger headline will inevitably grab more attention than body content that is smaller. This idea ensures consumers can locate the information they require quickly, improving their general experience. Understanding the goals of users and the tasks they are attempting to do is the first step in creating an efficient visual hierarchy. Designers must rank content according to the user's importance and relevancy. For example, the product image and price are usually shown more prominently on an e-commerce website since they are considered more significant than the in-depth product description. Additionally, good use of whitespace can improve visual hierarchy. Whitespace, also known as negative space, aids in the separation of various elements, resulting in a less cluttered and easier-to-read interface.

Another essential UI design guideline is consistency. Maintaining uniformity among all interface elements—such as buttons, navigation menus, fonts, and colors—is referred to as consistency. Because of the interface's consistency, users can better anticipate how things will operate and where information will be located. Using design systems or style guides, which offer specifications for the interface's functional and aesthetic elements, can help create consistency.

In addition to enhancing usability, a consistent design strengthens the company identification. A brand's personality and beliefs can be reflected in a unified style that designers develop using a consistent color palette, typeface, and iconography. A social media app might have a more lively and playful aesthetic, yet a financial app might have a more formal and professional design language. Interactions are likewise subject to consistency, guaranteeing that comparable actions yield comparable outcomes. Users will anticipate the same action elsewhere on the website, for example, if clicking a button on one section opens a new page.

A crucial component of UI design is color theory, which is the deliberate use of color to produce visually beautiful and helpful interfaces. In addition to communicating information and evoking feelings, colors can affect users' perceptions and actions. A basic understanding of color theory aids designers in selecting hues that complement one another and produce the intended effect.

A vital tool in color theory, the color wheel shows the connections between primary, secondary, and tertiary hues. When combined, complementary colors—opposite each other on the color wheel—can produce a striking appearance. On the color wheel, analogous hues are adjacent and can create a calming and beautiful effect. A balanced and dynamic appearance can be achieved using triadic color schemes, which use three colors uniformly distributed across the color wheel.

Designers need to consider not just harmonizing color schemes but also the psychological effects of colors. For example, blue is a common choice for banking and healthcare applications since it is frequently associated with calmness and confidence. Red is a powerful color for call-to-action buttons because it can convey excitement or urgency. Maintaining readability also requires ensuring that the text and background colors have enough contrast, especially for visually impaired users.

Another essential component of UI design is typography, which is the art and skill of organizing type. Good typography creates hierarchy, improves readability, and establishes the interface's mood. Making an intelligible and aesthetically pleasing design requires careful consideration of typeface and font size selection. Sans-serif fonts are viewed as modern and minimalist, while serif fonts, with their ornamental strokes, are typically associated with formality and tradition.

Hierarchy can be created in typography by altering the text's style, weight, and size. For instance, headings are typically bolded and larger than body text to emphasize their importance. Text elements such as captions and subheadings can be distinguished by size, weight, color, and spacing. Typographic consistency is also essential; an interface with many different types and styles may need to be more organized and attractive.

The space between text and other elements, or whitespace, is essential in typography. Enough whitespace makes a design balanced and uncluttered and enhances readability. Other crucial factors include letter spacing (tracking) and line spacing (leading). While suitable letter spacing improves the text's overall legibility, proper line spacing guarantees that text blocks are easy to read.

Effectively integrating typography and color theory can improve an interface's visual appeal and usefulness. One way to establish a clear visual hierarchy is to use a bold color for headers and a more neutral hue for body content. Similarly, a unified and harmonious appearance can be achieved by selecting a font that matches the overall color scheme. When using color and typography, designers must also take accessibility into account. They must ensure that text is readable by individuals with visual impairments and that color selections don't rely on color to convey information.

To sum up, the fundamentals of user interface design—visual hierarchy, consistency, color theory, and typography—are necessary to produce functional and aesthetically pleasing interfaces. Consistency fosters a unified and predictable experience, while visual hierarchy guarantees that users can quickly locate and concentrate on critical information. Typography improves legibility and establishes the interface's tone, while color theory aids designers in selecting hues that elicit the intended feelings and successfully communicate information. Designers can produce interfaces that are aesthetically pleasing, practical, and easy to use by comprehending and putting these ideas into practice. Together, these ideas help users navigate the UI and enhance the ease and enjoyment of their interactions.

Designing UI Elements

Designing UI elements such as buttons, icons, and navigation is a fundamental aspect of user interface design that significantly impacts the usability and aesthetics of a digital product. These elements guide users through the interface, helping them perform actions, understand information, and navigate between different sections. Responsive design considerations enhance the user experience by ensuring the interface adapts seamlessly to various devices and screen sizes. This section will explore the design principles for buttons, icons, and navigation and discuss the importance of responsive design.

Buttons are one of the most interactive elements in a user interface, acting as the primary means for users to act. Designing practical buttons involves considering their visual appearance, placement, and functionality. Visually, buttons should be distinct and easily recognizable. They should stand out from other elements on the screen, often achieved through color, size, and shape. The color of a button should contrast with the background to make it noticeable, and the use of consistent colors for buttons across the interface helps maintain uniformity. For instance, primary actions such as "Submit" or "Save" can be highlighted with a bold color, while secondary actions like "Cancel" might use a more subdued color.

The size and shape of buttons also play a crucial role in their effectiveness. Buttons should be large enough to be easily clickable, especially on touch devices where users interact with their fingers rather than cursors. Rounded corners are often preferred because they are perceived as more friendly and accessible for the eye to process. The placement of buttons is equally essential; primary buttons should be positioned where users can easily find and access them, usually at the bottom right of forms or dialogs, where users expect to see them.

Icons are another essential component of UI design, serving as visual shorthand for actions, objects, or concepts. Well-designed icons enhance the user experience by providing a quick, intuitive understanding of what an element represents. Icons should be simple and easily recognizable, avoiding unnecessary details that could confuse users. Consistency in icon design is critical; using a cohesive style throughout the interface helps users develop familiarity and predictability, which improves usability.

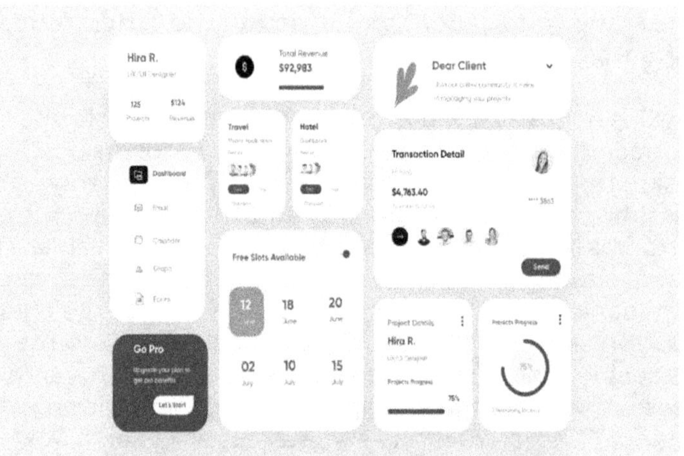

Icons should also be accompanied by text labels where necessary, particularly for complex or less common actions. This ensures clarity and reduces the cognitive load on users. For instance, a trash can icon typically represents a delete action, but pairing it with a "Delete" label removes ambiguity. When designing icons, it's essential to consider cultural differences and ensure that symbols are universally understood to avoid misinterpretation.

Navigation is the backbone of a user interface, allowing users to move between different product sections. Effective navigation design ensures that users can easily find what they are looking for and understand their current location within the interface. Various types of navigation patterns exist, including top navigation bars, side menus, and tab bars, each suited to different contexts and screen sizes.

Top navigation bars are commonly used for websites and desktop applications. They are positioned at the top of the screen, providing a horizontal list of links to the site's main sections. This type of navigation is intuitive and works well for interfaces with a limited number of main sections. Side menus, or slide-out menus, are often used for mobile applications where screen space is limited. These menus can be hidden by default and revealed when needed, preserving screen real estate for content. Tab bars, usually positioned at the bottom of the screen on mobile devices, provide quick access to the most frequently used sections of the app.

In addition to these navigation patterns, breadcrumb trails can help indicate the user's current location and their path to get there. This is particularly helpful in complex interfaces with multiple levels of hierarchy. Providing a search function is also essential for large interfaces, allowing users to quickly find specific content without navigating through various sections.

Responsive design is crucial for modern UI design, ensuring that the interface adapts seamlessly to different devices and screen sizes. With the proliferation of smartphones, tablets, and various screen sizes, designing interfaces that provide a consistent and optimized experience across all devices is essential. Responsive design involves using flexible grids, layouts, and media queries to adjust the design elements based on the screen size and resolution.

Several best practices should be followed when designing buttons, icons, and navigation for responsive interfaces. For buttons, ensuring that they remain touch-friendly on smaller screens is vital. This means maintaining an adequate size and spacing to prevent accidental clicks. Icons should be scalable, ensuring they stay clear and recognizable at different sizes. This might involve creating multiple versions of each icon optimized for other resolutions.

Navigation in responsive design often requires more significant adjustments. A top navigation bar might work well on larger screens, but on smaller screens, this could be replaced by a side menu or a hamburger menu to save space. It's essential to ensure that navigation remains easily accessible and intuitive regardless of the screen size. Collapsing less critical navigation items into a "More" menu can help maintain a clean and user-friendly interface.

Testing the responsive design on various devices is crucial to identify and address usability issues. This involves checking how the interface adapts to different screen sizes, orientations, and resolutions. Tools like responsive design mode in browsers or device emulators can aid this process, but testing on actual devices provides the most accurate results.

In conclusion, designing UI elements such as buttons, icons, and navigation requires careful consideration of their visual appearance, functionality, and placement. Practical buttons are easily recognizable, clickable, and positioned intuitively. Well-designed icons are simple, recognizable, and often paired with text labels to ensure clarity. Navigation must be intuitive and adaptable, providing a seamless experience across different interface sections. The responsive design further enhances the user experience by ensuring the interface adapts seamlessly to various devices and screen sizes. By following these principles and best practices, designers can create user interfaces that are visually appealing, highly functional, and accessible, providing a positive and consistent user experience across all devices.

Creating High-Fidelity Mockups

Making high-fidelity mockups is a crucial step in the UI design process that connects creative designs to the finished product. Stakeholders can see the finished product thanks to these realistic and thorough user interface mockups. Before entering the development phase, high-fidelity mockups are necessary to ensure the design satisfies user expectations and project objectives. Designers employ various UI design tools and meticulously prepare assets for development to do this.

Highly detailed designs that closely match the finished product are called high-fidelity mockups. Precise typography, colors, photos, layouts, and interactive features are all included. Instead of low-fidelity wireframes or sketches, high-fidelity mockups offer a thorough visual representation that enables stakeholders to understand precisely how the interface will appear and work. Making educated design decisions and obtaining accurate feedback require this degree of information.

To produce high-fidelity mockups, numerous tools are accessible, each with unique characteristics and functionalities. Nowadays, the most widely used tools for designers are Adobe XD, Sketch, Figma, and InVision. With its many tools for creating, prototyping, and collaboration, Adobe XD is a flexible tool. It is an excellent option for designers already familiar with Adobe's toolkit because of its connectivity with other Adobe Creative Cloud apps. Adobe XD is a good tool for producing intricate and dynamic mockups since it has powerful vector editing tools, responsive resizing options, and interactive prototype capabilities.

Another application that's very popular with macOS designers is Sketch. The Sketch is well-known for its user-friendly interface and vast plugin ecosystem, which make it simple for designers to produce high-fidelity mockups. Precise control over layouts and elements is made possible by its vector-based design tools, and the Symbols feature makes reusable components possible, guaranteeing consistency across the design. Sketch is a complete UI design solution because of its interaction with several prototype and collaboration tools, further improving its functionality.

Because of its collaborative features, the cloud-based design application Figma has become increasingly popular. Unlike traditional design tools, Figma is perfect for team-based projects since it enables numerous designers to work on the same project simultaneously. Figma has an extensive component system, interactive prototype capabilities, and various vector editing tools. Its real-time collaboration features facilitate smooth communication and feedback between team members, streamlining the design process. Furthermore, because Figma is browser-based, compatibility problems are eliminated, enabling designers to work from any device with an internet connection.

Another effective tool for producing interactive prototypes and high-fidelity mockups is InVision. With its design platform, InVision Studio, designers can create dynamic and captivating interfaces with its powerful animation and motion design capabilities. A variety of collaboration features are also offered by InVision, such as version control and commenting, which promote effective feedback and iteration. The InVision Design System Manager (DSM) contributes to consistency by enabling designers to establish and oversee shared design libraries.

After creating high-fidelity mockups, the next stage is getting the assets ready for development. To do this, design elements must be exported and optimized for implementation. Accurate asset preparation is essential to preserving the design's integrity and guaranteeing a seamless development transfer.

To begin the process of preparing assets, arrange the design files. According to designers, layers, and components should have explicit names and be arranged rationally. This arrangement makes it easier for developers to find specific elements and comprehend the design's structure. Consistent and effective asset management is made possible using symbols or components in programs like Sketch, Adobe XD, or Figma. Modifications made to one symbol instantly update all other instances of that symbol in the design.

Exporting assets entails choosing the design components that must be used and preserving them in the proper forms. UI assets are often stored in PNG, JPEG, SVG, and PDF formats. Because PNG supports transparency and offers lossless compression, it is frequently used for raster pictures and icons. JPEG does not support transparency, yet it is appropriate for high-quality photographs when the file size needs to be reduced. SVG is the recommended format for vector images and icons since it can be resized without sacrificing quality and is compatible with most current web browsers. Vector-based graphics that need exact scaling and high quality can be stored in PDF format.

Designers should give developers comprehensive specifications and guidance in addition to exporting materials. This contains details regarding dimensions, typography, colors, and spacing. This procedure is made more accessible by tools like Zeplin, Avocode, and the built-in features of Figma, which generate style guides and specifications straight from the design files. To ensure the execution follows the idea, these tools let developers download assets, examine CSS code snippets, and examine design aspects.

When creating assets, designers should also take responsiveness into account. This entails producing many asset versions for various screen sizes and resolutions. It is recommended to export photos and icons at different resolutions (1x, 2x, and 3x), for example, to make sure they look sharp on normal and high-density monitors. To guarantee a consistent user experience across devices, responsive design considerations also involve giving instructions for how the layout and elements should adjust to various screen sizes.

Designers and developers must work together and communicate during the asset preparation stage. Any problems or disagreements that surface from the design to the development phase are addressed with the support of frequent meetings and feedback sessions. Real-time communication is facilitated, and team alignment is ensured using collaboration tools and platforms like Microsoft Teams and Slack.

In summary, crucial phases in the UI design process include producing high-fidelity mockups and getting assets ready for production. High-fidelity mockups give a thorough and accurate depiction of the finished product, facilitating precise input and well-informed design choices. Several solid tools are available for developing these mockups and allowing collaboration, including Adobe XD, Sketch, Figma, and InVision. Design files must be arranged, components must be exported in the correct formats, thorough specifications must be provided, and responsiveness must be considered when developing assets. An excellent and unified user interface is ultimately the result of a smooth transition from design to development, which requires effective collaboration and communication between designers and developers.

CHAPTER VI

Setting Up the Development Environment

Choosing the Right Tools

Choosing the right tools is crucial for the success of software development projects. Integrated Development Environments (IDEs), Software Development Kits (SDKs), emulators, and version control systems play vital roles in the development process, enhancing efficiency, productivity, and collaboration. This section will delve into the importance of these tools and how they contribute to the development lifecycle.

Integrated Development Environments (IDEs) are software applications that provide comprehensive facilities to computer programmers for software development. An IDE typically consists of a source code editor, build automation tools, and a debugger. IDEs are designed to maximize developer productivity by providing tight-knit components with similar user interfaces. Some popular IDEs include Visual Studio, IntelliJ IDEA, Eclipse, and PyCharm.

Visual Studio, developed by Microsoft, is a powerful and versatile IDE widely used for developing applications in various programming languages, including C#, C++, Python, and JavaScript. It offers rich features like code completion, debugging tools, and integrated version control. Visual Studio also supports extensions, allowing developers to customize the IDE to their needs.

IntelliJ IDEA, developed by JetBrains, is another popular IDE, particularly favored for Java development. IntelliJ IDEA is renowned for intelligent code completion, deep static code analysis, and refactoring tools. It also supports various other languages and frameworks, making it a versatile choice for many developers.

Eclipse is a free, open-source IDE widely used for Java development, though it also supports other languages through plugins. Eclipse's modular architecture and extensive plugin ecosystem make it a highly customizable IDE that can be tailored to fit the specific needs of a project.

PyCharm, also developed by JetBrains, is a specialized IDE for Python development. PyCharm offers advanced code editing, debugging, testing features, and support for web development frameworks such as Django and Flask.

Software Development Kits (SDKs) are essential for developing applications for specific platforms or frameworks. SDKs provide the necessary tools, libraries, documentation, and sample code to help developers create software. For example, the Android SDK is essential for developing Android applications, providing APIs and tools to build, test, and debug apps for Android devices. Similarly, the iOS SDK is necessary for developing applications for Apple's iOS platform, offering tools like Xcode, Interface Builder, and a comprehensive set of APIs.

Using the appropriate SDK ensures that applications are built in compliance with the platform's standards and guidelines, which is crucial for performance, security, and user experience. SDKs also help streamline the development process by providing pre-built components and libraries, reducing the amount of code developers need to write from scratch.

Emulators are another essential tool in the development process, especially for mobile and embedded systems. Emulators mimic a specific device's hardware and software environment, allowing developers to test and debug their applications without needing physical devices. This is particularly useful in the early stages of development when access to all possible target devices could be more practical.

For instance, the Android Emulator is an essential tool for Android developers. It allows them to test their applications on various Android devices and configurations without needing the actual devices. The Android Emulator supports many features, including phone calls, text messaging, location services, and more, providing a comprehensive testing environment.

Similarly, the iOS Simulator, included with Xcode, enables iOS developers to test their apps on different iPhone and iPad models. The iOS Simulator provides a high-fidelity simulation of iOS devices, allowing developers to interact with their apps in a realistic environment.

Version control systems (VCS) are indispensable tools for managing changes to the source code over time. They allow multiple developers to work on the same project simultaneously, track changes, and revert to previous versions if needed. Git, Subversion (SVN), and Mercurial are the most widely used version control systems.

Git, developed by Linus Torvalds, is a distributed version control system that has become the de facto standard in the industry. Git allows developers to work on local repository copies, committing changes locally and then pushing them to a central repository. This distributed nature of Git makes it highly flexible and efficient, enabling parallel development and reducing the risk of conflicts.

Platforms like GitHub, GitLab, and Bitbucket provide cloud-based hosting for Git repositories, offering additional features such as issue tracking, pull requests, and continuous integration/continuous deployment (CI/CD) pipelines. These platforms facilitate collaboration and streamline the development workflow by integrating with other tools and services.

Subversion (SVN) is a centralized version control system widely used in the industry for many years. SVN maintains a central repository, with developers checking out copies of it and committing changes back to it. While SVN is not as flexible as Git, it is still a robust and reliable system for managing source code.

Mercurial is another distributed version control system that is similar to Git. Mercurial is known for its simplicity and performance, making it a suitable choice for projects of all sizes. It provides many of the same features as Git, including branching and merging, but focuses on ease of use.

Choosing the right version control system is crucial for the success of a development project. It helps manage the source code and facilitates collaboration, ensuring that all team members are working on the most recent version of the project. Version control systems also provide a history of changes, allowing teams to track progress, identify issues, and maintain accountability.

In conclusion, choosing the right tools—IDEs, SDKs, emulators, and version control systems—is essential for successful software development. IDEs provide comprehensive environments that enhance productivity and efficiency, while SDKs offer the necessary resources for developing platform-specific applications. Emulators enable developers to test and debug their applications in realistic environments without needing physical devices. Version control systems facilitate collaboration, manage changes to the source code, and provide a history of modifications. By selecting the appropriate tools for their needs, development teams can streamline their workflows, improve the quality of their software, and ultimately deliver better products.

Installing and Configuring Software

A fundamental stage in any software development project is installing and configuring software for development. This procedure includes configuring development environments and setting up Integrated Development Environments (IDEs) like Android Studio and Xcode to ensure effective coding, debugging, and testing. An environment that is stable, productive, and supports the many stages of the software development lifecycle can be created through proper installation and configuration.

Downloading the installer from the official Android developer website is the first step in configuring Android Studio, the official integrated development environment for Android. The installation process can begin by running the installer after it has been downloaded. Users are guided through the setup wizard, which lets them choose which components to install and the installation path. Installing the Android SDK and the Android Studio IDE simultaneously is advised during this procedure. Android Studio must be customized when the installation is finished.

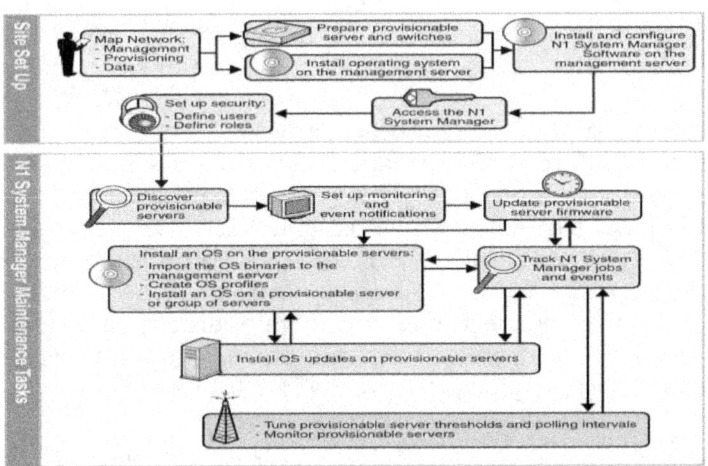

Users are invited to import settings from a previous installation or start again when they initially launch

Android Studio. The IDE then installs the most recent SDK components and searches for updates. Installing different versions of the Android SDK, system images, and other necessary tools is made possible by the Android SDK Manager, which is a crucial component of this configuration. Setting up emulators to test apps on various Android devices and screen sizes also requires configuring the Android Virtual Device (AVD) Manager.

Developers frequently alter Android Studio beyond the default configuration to suit their workflow. This involves setting the code editor with desired themes, fonts, and key bindings. Installing plugins can improve functionality, from version control integrations like Git to code linting and formatting tools. Another crucial step is configuring Gradle, the build automation tool used in Android development. Optimizing Gradle scripts guarantees effective dependency management and project builds.

The official IDE for developing iOS apps, Xcode, has a similar setup procedure. You may get Xcode from the Apple Developer website or the Mac App Store. After downloading, a setup wizard walks you through a simple installation. Additional components, including command line tools, need to be installed for Xcode after it has been installed. This can be done via Xcode's options.

Setting up the development environment in Xcode to support different iOS devices and simulators is part of the configuration process. Developers can manage and download simulator versions and iOS SDKs using the Xcode Preferences pane. To test apps on multiple devices, simulators for various iPhone and iPad models must be built and configured. In Xcode, developers may also create team accounts, which are required for code signing and distributing programs to real devices for testing.

In Xcode, customization goes beyond setting up the workspace. To increase efficiency, developers can change the editor's theme, install extensions, and modify the layout. Configuring Xcode also requires setting up build setups and schemes. By managing many build environments, including development, staging, and production, developers can ensure the app is built with the right resources and settings for each environment.

Customizing the larger development environment is crucial for a smooth workflow, even beyond setting up IDEs like Android Studio and Xcode. Installing version control programs like Git falls under this category. Installing Git entails downloading the installer and following the setup guidelines from the official Git website. For safe connection with remote repositories, Git must be configured with user details (name and email) and SSH keys set up after installation.

Another critical component of customizing the development environment is setting up dependencies and package managers. This frequently entails setting up Gradle scripts to manage project dependencies for Android development. Setting up CocoaPods or Swift Package Manager facilitates the management of third-party frameworks and libraries for iOS development. By making adding, updating, and removing dependencies easier, these tools guarantee that the project is constantly utilizing compatible and current libraries.

Configuring pipelines for Continuous Integration/Continuous Deployment (CI/CD) is another aspect of setting up development environments. DevelopingPrograms like GitHub Actions, CircleCI, and Jenkins automate developing, testing, and delivering applications. Configuring these tools entails setting up build, test, and deployment scripts to ensure that code changes are automatically tested and delivered to the proper environments. By minimizing manual involvement, this configuration not only speeds up development but also enhances the quality of the code.

Another technology that's frequently used to configure development environments is Docker, especially for web and backend development. Using Docker, developers can build containerized environments that are identical to the production environment, guaranteeing consistency between the testing, production, and development phases. Writing Docker files to provide the environment setup and using Docker Compose to manage multi-container apps are the two steps in setting up Docker. This configuration lessens the possibility of environment-specific problems by ensuring the program operates consistently across various settings.

Setting up Integrated Development Environments (IDEs) for particular programming languages and frameworks is another aspect of configuring development environments. For example, installing and configuring the Python interpreter, installing required plugins, and downloading and installing PyCharm from the JetBrains website are all part of setting it up for Python programming. Similarly, installing the IDE, configuring the Node.js runtime, and installing extensions like ESLint and Prettier for code formatting and linting are all necessary when setting up Visual Studio Code for JavaScript or Node.js development.

For development to be effective, debugging tools must be set up in addition to IDEs and dependencies. Setting up the Android Debug Bridge (ADB) enables developers to interact with Android devices to debug Android applications. Setting up the Xcode debugger for iOS development gives you access to debugging solid features like breakpoints, watchpoints, and stack traces. For web development, setting up remote debugging tools like Chrome DevTools is essential for identifying and resolving problems with web applications.

Another crucial factor to take into account while setting up development environments is security. This entails handling sensitive data, such as API keys and credentials, and establishing secure connections to databases and remote repositories. Sensitive data can be safely managed using tools like environment variable managers and secret management services, which ensure the data is not accessible in the codebase or version control systems.

Setting up communication and documentation tools is the final step in configuring development environments. Tools like Confluence, JIRA, and Slack enhance project management, documentation, and team communication. Setting up project boards, making documentation spaces, and combining communication channels with development tools are all part of configuring these technologies. This configuration enhances teamwork and output by ensuring that the development team is in sync and has access to all relevant data.

In summary, installing and setting up software for development is a thorough procedure that includes assembling IDEs such as Android Studio and Xcode, setting up development environments with required tools and dependencies, and streamlining the workflow to maximize effectiveness and productivity. Proper installation and configuration ensures a stable and productive environment for developers, enabling different stages of the software development lifecycle. Developers may increase code quality, optimize workflow, and produce better software by adequately configuring and setting up these tools.

Creating Your First Project

Creating their first project is an exciting turning point in any developer's career. Establishing a new app project and comprehending its structure are essential first stages that form the foundation for a productive development process. Choosing the appropriate tools, starting the project, and developing a thorough grasp of the different parts and files that comprise the project structure are all part of this process. Using Android Studio and Xcode as our primary examples for creating new app projects for iOS and Android, respectively, we will examine these features in detail here.

The first step in creating a new app project is selecting the right development environment. The official IDE for Android development is called Android Studio. It offers an extensive feature set and toolkit created especially for developing Android applications. Installing Android Studio requires downloading it from the official website. After installation, choose "Start a new Android Studio project" from the welcome screen when the IDE launches. Selecting a project name, location, and initial activity type are just a few processes the project creation wizard walks you through. An Android app's activities serve as the entry points for user interactions. Empty, Basic, and Fullscreen activities are examples of standard activity designs.

Similarly, Apple's official IDE for iOS programming is called Xcode. Xcode can be downloaded and installed from the Mac App Store. Launch Xcode after installation, then choose "Create a new Xcode project" from the welcome screen. Selecting a template for your project is one of the prompts provided by Xcode's project creation wizard. Examples of standard templates are Augmented Reality Apps, Games, and Apps. After choosing a template, you must provide information like the project name, organization ID, and language (Object-C or Swift). Xcode generates the project with a preset structure after these details are entered.

It is essential to comprehend the project structure to build and maintain it effectively. The project structure in Android Studio is separated into multiple files and directories. The "app" and "gradle" directories are the most significant. The application's materials and source code are in the "app" directory. The "src" folder, which is further subdivided into "main" and "test" folders, is situated in the "app" directory. The application code, XML layouts, Java/Kotlin files, and other resources are in the "main" directory. The "drawable" and "layout" folders in the "res" directory under "main" include picture assets; similarly, the "values" directory within "main" has XML files that define colors, strings, styles, and so on.

The project's "AndroidManifest.xml" file is essential in the "main" directory. It outlines hardware and software requirements, the app's package name, components (activities, services, broadcast receivers, and content providers), permissions, and other pertinent details. The Gradle build scripts needed to automate the application's development, testing, and deployment are in the "Gradle" directory. While the "build. gradle" file in the "app" directory specifies app-level settings, dependencies, and build configurations, the "build. gradle" file in the project root sets global project settings.

The project structure in Xcode is also arranged into multiple files and directories. A hierarchical view of the project files is available via the project navigator on the Xcode window's left side. "AppDelegate.swift," "ViewController.swift," and the "Main. storyboard" file are among the most significant directories. The application lifecycle functions are contained in the "AppDelegate.swift" file, which manages app-level events like starting, going into the background, and ending. The code for the first view controller, which controls the user interface for the application's main screen, is located in the "ViewController.swift" file. You can design and arrange the UI components of the application using a drag-and-drop interface in the "Main. storyboard" file, which is a visual depiction of the user interface.

A "Assets. xcassets" directory is also included in Xcode projects to handle picture assets and other resources. A vital configuration file called "Info. plist" holds information about the application, including its bundle identity, version number, supported interface orientations, and other variables. Localization files, launch screens, and icons are examples of extra resources that can be found in the "Supporting Files" subfolder.

Many tools are available for handling dependencies in Android Studio and Xcode. Using the Gradle build system, you can declare dependencies in the "build. gradle" files in Android Studio. Gradle will automatically download and include the libraries and frameworks your application needs if you provide them. Similarly, you can use Carthage, CocoaPods, or Swift Package Manager in Xcode to handle dependencies. These tools will obtain and integrate these libraries into your project, letting you specify dependencies in configuration files.

The next stage is to begin app development after the first project setup is finished and the project structure is known. Implementing the app's functionality usually entails writing Java or Kotlin code, defining layouts in XML, and generating activities and fragments in Android Studio. The user interface of an Android app is composed of activities. Users can engage with a single screen across each activity. Reusable user interface (UI) elements called fragments can be integrated inside activities to provide more adaptable and modular designs. Layout files employ XML to specify the appearance and organization of the user interface.

Writing Swift or Objective-C code, defining interfaces in Interface Builder, and developing view controllers are all part of the development process in Xcode. The UI of an iOS app is built around view controllers. A single screen or view hierarchy is controlled by each view controller. Xcode's integrated Interface Builder lets you graphically design and arrange UI elements. IBOutlet and IBAction let you attach UI elements to your code so you can manage user interactions and adjust the UI programmatically.

Throughout the development process, testing and debugging are essential components. For these tasks, Xcode and Android Studio both offer potent tools. The Android Emulator in Android Studio lets you test and run your app on simulated devices with various setups. A robust debugger with support for breakpoints, watches, and step-by-step execution is also included in the IDE. Similarly, you may test your app on many virtual iOS devices using the iOS Simulator included with Xcode. Xcode's debugger provides advanced functionality like breakpoints, variable examination, and memory analysis.

To summarize, developing a new app project and comprehending its structure is necessary to create your first project. Developers can initialize projects using established templates, specify essential settings, and efficiently arrange project files and directories using integrated development environments (IDEs) such as Android Studio and Xcode. Using the IDEs' tools, controlling dependencies, and comprehending the project structure are all essential elements in the development process. By grasping these fundamental concepts, developers can build reliable and maintainable apps and provide the groundwork for successful software development.

CHAPTER VII

Basic App Development

Writing Your First Lines of Code

Writing your first lines of code is an exciting milestone in learning programming languages. Understanding the basics is crucial whether you choose Java, Swift, or any other language. One of the most common traditions in the programming world is to write a "Hello World" application as the first step. This simple program outputs the phrase "Hello, World!" to the screen, introducing you to the syntax and structure of the language. In this section, we will explore the basics of programming languages through the creation of a "Hello World" app in both Java and Swift.

Java is one of the most widely used programming languages, known for its portability and robustness. It is an object-oriented language, which means it is based on the concept of objects that can contain both data and methods. To start with Java, you must install the Java Development Kit (JDK) and an Integrated Development Environment (IDE) like Eclipse or IntelliJ IDEA. Once your environment is set up, you can create a new project.

In Java, every application starts with a class definition. The class is the blueprint from which individual objects are created. Here is a simple "Hello World" program in Java:

```java
java public class HelloWorld { public static void main(String[] args) { System.out.println("Hello, World!");}}
```

This code defines a class named `HelloWorld`. The `main` method is the entry point of any Java application. The `public` keyword indicates that the method can be called anywhere. The `static` keyword means that the method belongs to the class rather than instances of the class. The `void` keyword signifies that the method does not return any value. The `main` method takes a single argument: an array of `String` objects. The line `System. Out.println("Hello, World!");` prints the text to the console. The `System` class provides access to system resources, and `Out` is a static member of this class representing the standard output stream.

On the other hand, Swift is a programming language Apple developed for iOS and macOS applications. It is known for its modern syntax, safety features, and performance. To start with Swift, you must install Xcode, the official IDE for developing Apple applications. Once Xcode is set up, you can create a new project and choose a template for a macOS or iOS application.

Here is how you can write a "Hello World" program in Swift: swift import Foundation print("Hello, World!")

The `import Foundation` statement imports the Foundation framework, which provides essential data types, collections, and operating system services to define your app's base layer of functionality. The `print` function is used to output text to the console. Unlike Java, Swift does not require a class definition or a primary method for this simple program, making it more concise and easier to read.

Both Java and Swift have their unique features and uses. Java's strength lies in its portability across different platforms, achieved through the Java Virtual Machine (JVM). Java code can run on any device with a JVM installed, making it a popular choice for enterprise applications and Android development. Swift, however, is specifically designed for Apple's ecosystem, offering seamless integration with macOS and iOS and providing powerful tools for building high-performance applications with a modern syntax.

When writing your first lines of code, understanding the basic structure and syntax of the language is essential. The "Hello World" program introduces these elements in both Java and Swift. In Java, you learn about classes, methods, and the importance of the primary method as the application's entry point. In Swift, you are introduced to the simplicity and power of the language's syntax and the use of built-in functions like `print.`

Programming languages also share some common concepts, such as variables, data types, operators, control flow, and functions. Variables are used to store data that the program can manipulate. Data types define the data that can be stored, such as integers, floating-point numbers, and strings. Operators are used to perform operations on variables and values. Control flow statements like loops and conditionals determine the flow of execution based on certain conditions. Functions are reusable blocks of code that perform a specific task.

In Java, variables are declared with a specific data type, for example, Java int number = 5; String text = "Hello".

In Swift, variables are declared with the `var` keyword, and the compiler can infer the data type: swift var number = 5 var text = "Hello"

Both languages provide control flow statements such as `if,` `else,` `for,` and `while.` Here is an example of a simple conditional statement in Java: Java if (number > 0) {System. Out.println("The number is positive.").

} else {System. out.println("The number is negative or zero.");}

And in Swift: swift if number > 0 {print("The number is positive.")} else {print("The number is negative or zero.")

As you continue to write more complex programs, you will encounter concepts such as object-oriented programming (OOP), which is fundamental to Java. OOP involves creating classes and objects, encapsulating data, and using inheritance and polymorphism to build flexible and reusable code. Swift also supports OOP but strongly emphasizes protocol-oriented programming (POP), which uses protocols to define interfaces and provides a more flexible approach to code reuse.

In conclusion, writing your first lines of code in Java or Swift is essential in learning to program. The "Hello World" program introduces you to the basic syntax and structure of the language. Understanding these fundamentals lays the groundwork for more complex programming tasks. Whether you choose Java for its portability and robustness or Swift for its modern syntax and integration with Apple's ecosystem, mastering the basics will set you on the path to becoming a proficient programmer. As you progress, you will discover the power and flexibility of programming languages, enabling you to create various applications and solve real-world problems.

Building Core Features

Building core features of an application involves creating a user interface (UI) that allows users to interact with the app, handling user input effectively, and providing smooth navigation between different screens or sections. This process is fundamental to creating a functional and user-friendly application, whether you are developing for the web, mobile, or desktop platforms. This section will explore the critical aspects of adding UI elements, handling user input, and implementing navigation in an application.

UI elements are the building blocks of any application's interface. They include buttons, text fields, images, labels, and more complex components like tables and sliders. Adding these elements to your application involves using a UI framework or toolkit provided by your development environment. For instance, HTML, CSS, and JavaScript are used in web development to create and style UI elements. In mobile development, platforms like Android and iOS offer their UI frameworks: Android uses XML layouts and the View system, while iOS uses Interface Builder with UIKit or SwiftUI.

To illustrate, let's consider adding a button to a mobile application. In Android, you would define the Button in an XML layout file: XML <Button

android:id="@+id/button"

android:layout_width="wrap_content"

android:layout_height="wrap_content"

android:text= "Click Me"/>

In iOS, using SwiftUI, you would define the Button in a Swift file: swift Button (action: { print("Button clicked")}) {Text ("Click Me")}

Once the UI elements are in place, handling user input is the next critical step. User input can come in various forms, such as tapping a button, entering Text in a field, or selecting an item from a list. Handling this input requires setting up event listeners or handlers that respond to user actions.

For instance, in an Android application, to handle a button click, you would set an `OnClickListener` in your activity or fragment: javaButton myButton = findViewById(R.id.button).

myButton.setOnClickListener(new View.OnClickListener() {

@Override public void onClick(View v) {// Handle button click Toast.make text(getApplicationContext(), "Button clicked", Toast.LENGTH_SHORT).show();}}).

In SwiftUI for iOS, handling a button click is integrated into the `Button` declaration:

swift Button(action: {print("Button clicked")}) {Text("Click Me")}

Handling text input from a user involves setting up a text field and a listener for text changes. In an Android application, you can use an `EditText` view along with a `TextWatcher`: java EditText myEditText = findViewById(R.id.myEditText); myEditText.addTextChangedListener(new TextWatcher() {

@Override public void context changed(CharSequence s, int start, int before, int count) {// Handle text change}

@Override public void before exchanged(CharSequence s, int start, int count, int after) { // Do something before text is changed }

@Override public void after exchanged(Editable s) {// Do something after text is changed}}).

In SwiftUI, you use a `TextField` and bind it to a state variable: swift @State private var text input: String =" var body: some View {TextField("Enter text," Text: $textInput).padding()}

Effective navigation within an application ensures that users can quickly move between different screens or sections. This involves designing a logical flow and using navigation components provided by the platform. In Android, navigation is typically managed using activities and fragments. You can use explicit intents to start a new activity or the `FragmentManager` to replace fragments.

For example, to navigate to a new activity in Android, you might use the following code:

java Intent = new Intent(CurrentActivity.this, NewActivity.class); startActivity(intent).

In iOS, navigation is often managed by `UINavigationController` when using UIKit or by navigation links in SwiftUI. Here is how you might navigate to a new view in SwiftUI: swift NavigationLink(destination: NewView()) {Text ("Go to New View")}

For more complex navigation scenarios, such as deep linking or handling backstacks, both Android and iOS provide advanced navigation libraries and patterns. Android Jetpack's Navigation component simplifies implementing navigation, especially for apps with multiple fragments and nested navigation. Similarly, SwiftUI's navigation system allows for complex navigational structures using declarative syntax.

Creating a cohesive user experience involves adding UI elements, handling user input, and considering accessibility and responsiveness. Accessibility ensures that people with disabilities can use your application. This consists of using semantic elements, providing alternative Text for images, and ensuring that all interactive elements are reachable via keyboard and screen readers.

Responsiveness ensures your application looks and works well on different screen sizes and orientations. This can involve using flexible layouts, responsive design techniques, and testing on various devices. CSS media queries and frameworks like Bootstrap help achieve responsiveness in web development. In mobile development, constraint-based layouts in Android and Auto Layout in iOS help create adaptive interfaces.

To summarize, building an application's core features involves adding UI elements, handling user input, and implementing smooth navigation. You can create interactive and engaging applications by adding buttons, text fields, and other UI components. Handling user input effectively involves setting up event listeners and managing the input data. Navigation ensures users can move seamlessly between different application parts, enhancing the overall user experience. By focusing on these core aspects and considering accessibility and responsiveness, you can build applications that are not only functional but also user-friendly and inclusive.

Debugging and Troubleshooting

Troubleshooting and debugging are vital abilities for any programmer to possess. The development process involves more than just writing code; it also ensures it operates correctly and efficiently. Debugging is finding, locating, and repairing mistakes or defects in the code, whereas troubleshooting deals with more general problems that could come up during the software development process. This section explores typical programming mistakes, how to solve them, and the instruments and methods for debugging.

Syntax errors are among the most prevalent mistakes made when programming. When the code violates the conventions of the programming language, syntax errors arise. Because most integrated development environments (IDEs) reveal these problems immediately, they are frequently challenging to find. For example, a syntax error occurs when a semicolon is omitted at the end of a statement in Java or when parentheses are not matched in Python. Correcting the syntax by the language requirements is typically required to fix these mistakes. Nonetheless, since error messages frequently direct the reader to the offending line of code, it's imperative that you carefully study them.

Runtime errors are another common problem that arises during program execution. These errors are more challenging to diagnose because they result from several situations, like dividing by zero, accessing illegal memory locations, or attempting to utilize an object that hasn't been initialized. The program's execution may be interrupted by a `TypeError` in Python or a `NullPointerException` in Java. Programmers must follow the flow of the program's execution and check the state of variables at various stages to fix runtime issues. Using debugging tools to set breakpoints and step through the code to see where things go wrong are common steps in this process.

Another kind of bug that can be very sneaky is a logical error. Logical mistakes do not cause the application to crash or display error messages, unlike syntax or runtime problems. Instead, they lead to improper software operation. For example, if a loop calculates the sum of numbers, it can use the wrong variable and get an inaccurate total. Thorough testing and verification of the program's output against expected outcomes are necessary to detect logical mistakes. One efficient method is using unit tests, which are automated tests that confirm the behavior of distinct program components. Programmers can identify logical flaws early in the development process by creating test cases that check for different inputs and their expected outputs.

Errors must be found and fixed with the use of debugging tools. Most contemporary IDEs provide powerful debugging tools that let programmers examine variables, step through their code, and assess expressions. For example, developers can investigate the application's state by pausing the program's execution at specific lines of code using breakpoints, which they can define in Visual Studio. Then, they can walk over functions to bypass parts of code that are not urgently concerning or step into them to investigate their behavior.

The usage of logging is another effective debugging method. Developers can send messages to a log file or console by sprinkling logging statements throughout the code. This gives them insight into the program's state and flow of execution. Since logging offers developers access to the sequence of events that preceded the error, it is especially helpful in identifying problems that are hard to replicate. For customizing and managing log output, libraries such as the `logging` module in Python or Log4j for Java offer many options.

Tools such as profilers and debuggers are priceless for more complex debugging. Developers can examine the program at a shallow level using debuggers, such as GDB for C/C++ or PDB for Python, to examine register values, memory contents, and other aspects. Profilers, on the other hand, measure the majority of the program's time spent to help discover performance bottlenecks. Developers can optimize their code for better speed using tools like Python's `cProfile` module's built-in profiler or Java's JProfiler, which can offer thorough information on function calls, execution times, and memory usage.

Comprehending and managing exceptions is an essential component of debugging. Errors that arise while a program is being executed can be handled using exceptions. Languages like Python and Java provide robust exception-handling frameworks, enabling developers to identify and elegantly manage failures. Using {try-catch` blocks in Java or {try-except} blocks in Python, programmers can catch exceptions, record pertinent data, and execute remedial measures without the program crashing. Building robust apps that can recover from unforeseen circumstances and carry on with operations requires proper exception handling.

Apart from utilizing these technical instruments and methods, creating a systematic approach to debugging is crucial. This entails dissecting the issue, formulating plausible causes, and testing these theories via experimentation. A thorough understanding of the codebase, its operating environment, and the relationships between its components is frequently necessary for effective debugging. Collaborative debugging, in which developers diagnose and resolve problems with one another, can also be very successful because it applies a variety of viewpoints and levels of skill to the problem.

Using automated testing frameworks is another cutting-edge debugging strategy. Every time the code is changed, automated tests can run a battery of tests to ensure no new defects are discovered, and no functionality is broken. These tests can be automated by continuous integration (CI) tools like Jenkins or GitHub Actions, giving developers fast feedback. Teams can reduce the overall cost and labor of debugging by catching and fixing errors early on by integrating automated testing into the development cycle.

To sum up, troubleshooting and debugging are essential steps in the software development process. Producing dependable software requires understanding typical defects, including syntax, runtime, and logical errors, as well as knowing how to remedy them. Various debugging tools and approaches, such as sophisticated debuggers and profilers or simple breakpoints and logging, enable effective problem diagnosis and resolution. Developers can maintain and improve code quality with the help of automated testing, efficient exception handling, and a systematic approach to debugging. Programmers may create reliable, effective apps that satisfy user needs and endure the rigors of real-world use by learning these abilities.

CHAPTER VIII

Advanced Features and Integrations

Implementing Backend Services

An essential component of developing web and mobile applications is implementing backend services. Business logic, database transactions, user authentication, and API-based connection with other services are all managed by an application's backend. Understanding backend development is essential to understanding database administration, API integration, and server-side programming. In addition to introducing backend programming, this section looks at databases and APIs and how they can be utilized to create scalable, reliable systems.

Choosing a programming language and server-side development framework is the first step in backend development. Python, Ruby, Java, PHP, and JavaScript (with Node.js) are popular languages for backend programming. Building web apps is made more accessible by frameworks specific to each language. For example, Python frequently uses Django or Flask, Ruby depends on Ruby on Rails, Node.js uses Express.js, and Java frequently uses Spring Boot. These frameworks offer an organized approach to middleware integration, routing management, and HTTP request handling.

Data management is a fundamental aspect of backend development. This entails storing, retrieving, and manipulating data via databases. Relational and non-relational (NoSQL) databases are the two general categories into which databases fall. Information is arranged into tables with rows and columns using relational databases like Oracle, PostgreSQL, and MySQL.

They manage and query data using Structured Query Language (SQL). Applications requiring complex transactions and data integrity are best suited for relational databases.

NoSQL databases, on the other hand, like MongoDB, CouchDB, and Cassandra, store data in a variety of formats, including documents, graphs, JSON, and key-value pairs. NoSQL databases, such as social networking platforms, content management systems, and real-time analytics, are frequently utilized in applications that must expand horizontally. These databases are built to handle massive volumes of unstructured or semi-structured data.

Setting up a database server, creating the database schema, and utilizing an Object-Relational Mapping (ORM) tool to communicate with the database programmatically are usually involved in integrating a database into a backend application. By mapping database tables to application classes, an ORM library enables developers to write database queries in a programming language rather than in raw SQL—for instance, the Python ORM Django, the Java Hibernate, and the Node.js Sequelize.JavaScript is popular ORM utilities that make database interactions more efficient.

Creating and using APIs (Application Programming Interfaces) is a crucial component of backend development. APIs lay out a set of guidelines and procedures for using the services provided by the application. They provide communication across disparate software systems, making it possible to integrate mobile apps, web frontends, and third-party services with the backend.

RESTful (Representational State Transfer) services or GraphQL endpoints are the most common ways APIs are implemented. REST is an architectural style that carries out CRUD (Create, Read, Update, Delete) operations on resources designated by URLs using standard HTTP methods like GET, POST, PUT, and DELETE. An application for a blog, for instance, may have a RESTful API with endpoints like `/posts} to get all blog entries, `/posts/{id}} to get a particular article, and `/posts} with an article request to add a new post.

Facebook's GraphQL provides a more versatile method for data querying. GraphQL reduces the quantity of data carried over the network and improves performance by offering a single endpoint where clients can specify the precise data they require in place of several endpoints. For example, in contrast to REST, which may require many endpoints to get the same data, a GraphQL query for a blog application can request the title, content, and author details of a specific post in a single request.

Developers utilize the framework of their choice to build routes, manage HTTP requests, and return responses in a structured format, such as JSON, to implement APIs. The request-handling pipeline can be extended with middleware functions to perform input validation, logging, and authentication. For instance, Express.js middleware functions in a Node.js application can log request information for monitoring and debugging or verify if a user is authenticated before granting access to specific routes.

The permission and authentication processes are essential to backend services. Through methods like JWT (JSON Web Tokens), OAuth tokens, and username-password combinations, authentication confirms the identity of users. What resources and functions authenticated users can access depends on their authorization level. Ensuring that only authorized users can execute specific tasks and safeguarding sensitive data requires implementing secure authentication and authorization methods.

Backend services must frequently integrate with external APIs and manage data and user interactions. These integrations can offer additional features like email notifications, social network sharing, and payment processing. An application's functionality can be significantly increased by utilizing APIs from providers like Firebase for real-time databases and authentication, Twilio for messaging, and Stripe for payments.

For instance, connecting a payment gateway like Stripe requires handling webhooks to get alerts of payment events, creating payment intents through API calls, and securely storing client data. Similarly, establishing email templates, setting up API keys, and sending transactional emails over the API are necessary when integrating with an email provider such as SendGrid.

In backend development, scalability and performance are essential factors to consider. The backend needs to expand load without compromising performance as the number of users and data grows. This entails dividing the load among several servers or microservices, optimizing database queries, and implementing caching techniques. Commonly utilized to accomplish these goals are containerization technologies like Docker and Kubernetes for building scalable microservices architectures, load balancers for spreading traffic, and Redis for caching.

To keep backend services operating smoothly, monitoring and logging are also necessary. Tools for monitoring applications such as Prometheus, Grafana, and Datadog offer up-to-date information about an application's functionality, resource use, and possible problems. Using logging frameworks, developers may quickly detect and fix issues by gathering comprehensive information about the application's behavior.

To sum up, putting backend services into place is complex and includes server-side programming, database administration, API creation, and integration with other services. Developers can build robust and scalable backend services by selecting the appropriate programming language and framework, creating an effective database schema, and implementing reliable APIs. Building dependable applications requires careful consideration of third-party API usage, secure authentication and authorization methods, and performance and scalability optimization techniques. Developers may guarantee that their backend services satisfy the application's and users' changing needs using monitoring, logging, and continual optimization.

Adding Interactive Elements

Adding interactive elements to an application enhances user engagement and improves the overall user experience. Among these elements, multimedia components like images, audio, video, animations, and transitions significantly make applications more dynamic and appealing. Implementing animations and transitions requires a combination of design principles and technical skills to ensure that they look good and perform efficiently. This section explores the process of working with multimedia, focusing on the implementation of animations and transitions in modern applications.

Multimedia elements are essential for creating rich, interactive experiences in applications. Images, audio, and video can convey information more effectively than text, making applications more intuitive and engaging. Integrating multimedia into an application involves using various tools and libraries that handle the rendering and playback of these elements. For instance, HTML5 provides native support for embedding audio and video in web pages, while mobile development platforms like Android and iOS offer APIs for handling multimedia content.

Implementing animations involves creating visual effects that bring elements to life. Animations can range from simple effects like fading and sliding to complex sequences that affect multiple aspects and transitions. The primary goal of animations is to enhance user interaction by providing visual feedback and improving the flow of the application. For instance, animating a button when clicked can give users a sense of action and response, making the interaction more satisfying.

CSS (Cascading Style Sheets) is a powerful animation tool in web development. CSS animations allow developers to animate most HTML elements without requiring JavaScript. Keyframes define the stages of an animation, specifying the properties of a component at various points in time. For example, a simple animation to change the opacity of an element from 0 to 1 can be defined using CSS keyframes: C" @keyframes fadeIn { from { opacity: 0; } to { opacity: 1; }}.element {animation: fadeIn 2s ease-in-out;}

This CSS code creates a fade-in effect that takes two seconds to complete. The `ease-in-out` timing function provides a smooth transition, starting slowly, speeding up, and then slowing down again towards the end.

JavaScript is often used to control animations dynamically and create more complex effects. Libraries like jQuery, GreenSock (GSAP), and anime.js provide extensive functionality for animating elements on the web. These libraries simplify creating, controlling, and sequencing animations, allowing developers to focus on the creative aspects rather than the technical details.

For example, using GSAP to animate the position and scale of an element could look like this: javas "it gap. to(".element," { duration: 1, x: 100, scale: 1.5, ease: "power2.inOut"}).

This code snippet moves the element with the class `.element` 100 pixels to the right and scales it up 1.5 times over one second, using an easing function for a smooth transition.

Android and iOS in mobile development provide robust frameworks for implementing animations. On Android, the `Animator` and `Animation` classes allow developers to animate properties of UI elements. For example, to animate a view's view and translation, you can use the `ObjectAnimator` class: java Obje" Animator fadeIn = ObjectAnimator.ofFloat(view, "alpha," 0f, 1f); fadeIn.setDuration(1000).

ObjectAnimator moveUp = ObjectAnimator.ofFloat(view, "translationY", 100f, 0f).

moveUp.setDuration(1000).

AnimatorSet = new AnimatorSet().

animatorSet.playTogether(fadeIn, moveUp).

animatorSet. Start();

This code creates two animations: one for fading in the view and another for moving it upward. The `AnimatorSet` class is used to play both animations simultaneously.

In iOS, the `UIView.animate` method provides a simple way to create animations. For example, to animate the opacity and position of a view, you can use the following Swift code:

swift UIView."imate(withDuration: 1.0) {view.alpha = 1.0

view.transform = CGAffineTransform(translation: 0, y: -100)}

This code fades in the view and moves it up by 100 points over one second.

Transitions are another vital aspect of interactive design. They provide smooth changes between different states or screens in an application, enhancing the overall user experience. Transitions can be used to navigate between pages, switch views, or change the state of UI elements. In web development, CSS transitions allow developers to change property values smoothly over a specified duration. For example, to create a transition effect for changing the background color of an element, you can use the following CSS:

```css
.element {transition: background-color 0.5s ease-in-out;}.element:hover {background-color: blue;}
```

When hovered over, this code changes the element's background color to blue, taking 0.5 seconds to complete the transition.

In mobile development, transitions between screens are handled by the navigation framework of the platform. On Android, the `Fragment` and `Activity` classes provide methods to define custom transitions. For instance, you can use the `FragmentTransaction` class to set enter and exit animations for fragment transitions:

```java
FragmentTransaction transaction = getSupportFragmentManager().beginTransaction().

transaction.setCustomAnimations(R.anim.enter, R.anim.exit, R.anim.pop_enter, R.anim.pop_exit).

transaction.replace(R.id.container, new fragment).

transaction.addToBackStack(null).

transaction.commit().
```

This code sets custom animations for fragment transitions, specifying animations for entering, exiting, and popping the backstack.

In iOS, view controllers handle screen transitions. Using the `UIViewController` class, you can define custom transition animations. For example, to present a view controller with custom animation, you can use the following Swift code:

swift let new view controller = NewViewController()

new view controller.modalTransitionStyle = .crossDissolve present(new view controller, animated: true, completion: nil)

This code sets a cross-dissolve transition style for presenting the new view controller.

In conclusion, adding interactive elements through multimedia, animations, and transitions significantly enhances the user experience of applications. Whether a web or mobile application, using CSS, JavaScript libraries, and platform-specific frameworks enables developers to create visually appealing and responsive interfaces. Animations provide feedback and improve interaction flow, while transitions ensure smooth navigation between different states or screens. By mastering these techniques, developers can build applications that are not only functional but also engaging and delightful to use.

Integrating Third-Party Services

Integrating third-party services can significantly improve an application's usability and functionality. Payment gateways, social networking networks, and other APIs offer robust functionality to reduce development time and give users dependable, well-known tools. But integrating these services calls for thorough preparation, adherence to industry best practices, and consideration of security and performance issues. To better understand how to integrate third-party services, this section will concentrate on social networking and payment channels.

Applications can use the user base and popularity of social media platforms such as Instagram, Twitter, and Facebook by integrating them with other services. Features like content sharing, social feed embedding, and social login are examples of these connections. Social login, which enables users to connect to an application using their pre-existing social media identities, is one of the most popular integrations. This lowers friction, streamlines the registration process, and boosts user engagement.

Developers usually utilize OAuth, an open standard for access delegation, to implement social login. Applications can safely access users' data with OAuth without disclosing their login credentials. For instance, setting up the OAuth redirect URIs, registering the application with Facebook, and getting an app ID and secret are all necessary steps in integrating Facebook login. Once set up, users can be redirected by the program to Facebook for authentication. Facebook will then provide an access token, which the application can use to obtain the user's profile data.

Developers must ensure shared material adheres to platform requirements and is structured correctly when adding social media sharing tools. This usually entails tracking interaction and producing shareable content using the platform's SDKs or APIs. For instance, Twitter offers an SDK in JavaScript that enables developers to make tweet buttons, distribute content, and incorporate tweets into applications. Similarly, sharing posts, images, and videos straight from an application is made possible by Facebook's Graph API.

Using the platform's widgets or APIs to show content dynamically is necessary when embedding social media feeds. This can involve showing the postings from a Facebook page, a Twitter timeline, or an Instagram user's images. These integrations frequently entail obtaining data via the platform's API and utilizing HTML, CSS, and JavaScript to present it inside the application. To avoid throttling or temporary bans, it is crucial to manage API rate limitations and ensure the application doesn't send out more requests than permitted.

Payment gateway integration is another essential component of many apps, particularly e-commerce platforms. Online purchases may be made safely and efficiently with the help of payment gateways like Square, PayPal, and Stripe. Developers have three things to consider when integrating a payment gateway: security, user experience, and financial regulatory compliance.

Choosing the best supplier based on the application's requirements is one of the first stages in integrating a payment gateway. Transaction costs accepted payment options, accessibility by region, and simplicity of integration are all critical factors to consider. For example, due to its comprehensive documentation and developer-friendly API, Stripe is a well-liked option for numerous developers.

Setting up the required configurations, including making an account, getting API keys, and configuring webhooks, comes after choosing a payment gateway. Real-time notifications concerning events like successful payments, unsuccessful transactions, and refunds are obtained through webhooks. If webhooks are handled correctly, the program can react to these events, such as updating order statuses and sending confirmation emails.

Creating a safe and easy-to-use checkout process is essential when implementing the payment process. This entails protecting payment information by adhering to PCI-DSS (Payment Card Industry Data Security Standard) requirements, verifying user inputs to prevent fraud, and encrypting data transferred between the client and server using SSL/TLS. Many payment gateways offer hosted payment pages or pre-built user interface components to make this procedure easier and guarantee security standard compliance.

User feedback and error handling are crucial elements of a flawless payment process. When a payment fails, users should receive concise and helpful messages outlining the reason for the failure and instructions on how to fix it. For instance, the application should notify the user and offer substitute payment options if a payment is rejected because there aren't enough funds.

Developers can incorporate features like multi-currency support, transaction history, and payment method save-for-later to improve the customer experience further. These attributes can potentially boost user comfort and confidence, raising conversion rates and boosting client happiness.

An application can gain tremendous value by integrating third-party services in addition to social media and payment channels. Examples include using analytics tools like Google Analytics or Mixpanel to track user behavior and application performance, integrating cloud storage solutions like AWS S3 or Google Cloud Storage for managing files and media, and integrating email services like SendGrid or Mailchimp for email notifications and marketing campaigns.

It is imperative to consider the integration's long-term maintainability and scalability when integrating any third-party service. This entails updating dependencies, monitoring the service's API modifications, and building the integration modular and expandable. To simplify the process of switching providers or updating the integration without requiring substantial code changes, the application logic can be separated from the specifics of the service's implementation by utilizing an abstraction layer or wrapper around the third-party API.

Integrating third-party services requires careful consideration of security issues. Developers must safeguard confidential information and make sure that their application is impervious to assaults like cross-site scripting (XSS), man-in-the-middle (MITM) attacks, and cross-site request forgery (CSRF). This entails adhering to best practices, which include sanitizing and verifying inputs, utilizing HTTPS for all connections, and putting strong authentication and authorization procedures in place.

Performance is yet another critical factor. Third-party service integration may result in extra latency and other failure points. To reduce the impact on the speed of the application, developers should employ asynchronous processing whenever possible, optimize API queries, and incorporate caching mechanisms. Monitoring and logging are crucial to identifying and treating problems with third-party integrations. Sentry, New Relic, and Datadog are a few examples of tools that offer real-time monitoring, error tracking, and performance insights to developers so they can swiftly discover and fix issues.

In conclusion, integrating third-party services like social media sites and payment gateways can significantly improve an application's usability and functionality. However, it necessitates thorough planning, adherence to industry best practices, and consideration of performance, security, and maintainability. Through careful provider selection, configuration of integrations, and adherence to best practices for safe and effective deployment, developers may harness the potential of third-party services to create dependable, expandable, and intuitive apps.

CHAPTER IX

Testing and Optimization

Types of Testing

Testing is crucial in the software development lifecycle because it guarantees that programs satisfy specifications, work as intended, and offer a positive user experience. Different testing methodologies focus on various facets of software quality. Among the most crucial are unit, integration, and user interface testing; each has a distinct function in the validation process. Additionally, testing can be done manually or automatically, with each method offering pros and cons. These testing methods are examined in this section, along with the differences between automated and manual testing.

The most detailed kind of testing, unit testing, concentrates on specific features or parts of an application. Unit testing aims to confirm that each software unit operates as intended when used separately. This entails creating tests that verify that functions or methods perform as designed for a range of inputs by comparing their output to the expected results. Developers usually use testing frameworks like JUnit for Java, NUnit for .NET, or pytest for Python to write unit tests. These tests help find flaws early in the development process because they are typically quick to run and give instant feedback on whether the code is correct. Developers may locate a breakdown precisely by isolating units, which makes maintenance and debugging easier.

Conversely, integration testing looks at how various parts or units of an application work together. Integration testing verifies that components function as intended when combined, whereas unit testing ensures that components operate appropriately when separated. This testing is essential for locating problems that result from integrating various modules, like mismatched interfaces, inconsistent data formats, or unexpected behavior brought on by component interactions. Setting up a test environment that resembles the production environment may be necessary for integration tests, which are more complex than unit tests. Integration testing is generally performed using testing frameworks such as Mocha for JavaScript and TestNG for Java. Good integration testing ensures that the various parts of the application function together nicely and offers a more thorough evaluation of the system's behavior.

User interface testing, or UI testing, is the process of confirming that an application's graphical user interface (GUI) satisfies the requirements and offers a satisfactory user experience. This entails verifying that the visual components—such as buttons, menus, and forms—work properly and display effectively on various hardware and web browser combinations. Validating user behaviors, such as typing, clicking, and navigating the program, is another aspect of user interface testing. UI testing is frequently automated with tools like XCTest for iOS apps, Appium for mobile applications, and Selenium for web applications. To ensure that users can interact with the application as intended and that the UI reacts to user activities correctly, user interface (UI) testing is crucial, even though it can be more time-consuming and complex than unit or integration testing.

The two methods for carrying out tests are automated testing and manual testing. When testing manually, human testers carry out test cases without using automated tools. Testers record the outcomes after completing a predetermined set of tasks. This method works exceptionally well for exploratory testing, in which testers investigate the program to find possible problems that should have been included in the test plan. For usability testing, where human judgment is needed to evaluate the user experience, manual testing is also beneficial. However, manual testing can be less effective, time-consuming, and prone to human for large-scale or recurrent testing mistakes.

In contrast, automated testing uses software tools to test cases automatically. Computerized tests are perfect for regression testing because they can be run repeatedly without human intervention. Regression testing requires running the same tests after every code change to ensure that the functionality already in place is not affected. Running huge test suites fast and reliably is made possible by automated testing, which is very effective. For various application types, automated tests can be created and run with the help of tools such as Selenium, JUnit, and Pytest. Automated testing can significantly reduce long-term testing work and increase process reliability, but it does need an initial investment in building test scripts and setting up test infrastructure.

Every testing method, whether automated or manual, has advantages and disadvantages. Because manual testing is adaptable and quick to adjust, it may be used for exploratory testing, testing new features, and evaluating the user experience. It is slower, more labor-intensive, and might need to be more practical for repetitive or large-scale testing operations. In contrast, large test suites can be executed with automated testing faster than manual testing due to its efficiency, consistency, and scalability. Regression testing, performance testing, and situations requiring repeated runs of the same tests are areas in which it excels. Automated testing, however, may be less helpful in testing dynamic user interactions or carrying out exploratory testing because it takes up-front work to create and maintain test scripts.

In reality, the most effective strategy to guarantee software quality and attain thorough test coverage is frequently a mix of automated and manual testing. This hybrid technique combines the best features of both approaches by utilizing automated tests for repetitive and extensive testing tasks and manual testing for exploratory, usability, and new feature testing. Development teams can ensure that their apps are extensively tested, that problems are found and fixed quickly, and that the user experience is maximized by combining the two methods.

In summary, while unit testing, integration testing, and user interface testing focus on various facets of software quality and operation, they are all essential components of the software testing process. While UI testing verifies the user interface and interactions, integration testing looks at how components interact. Unit testing concentrates on individual components. There are two different ways to run tests: manually and automatically. Each has benefits and drawbacks of its own. To achieve thorough test coverage and guarantee the overall quality of the product, a combination of automated and manual testing is frequently the best approach. Development teams can create dependable, dependable, and easy-to-use apps by comprehending and implementing these different testing types and methodologies.

Performance Optimization

Performance optimization is crucial for enhancing an application's speed, responsiveness, and overall user experience. In a competitive digital landscape, users expect apps to be fast and efficient, and any lag or delay can lead to dissatisfaction and abandonment. Key focus areas for performance optimization include improving app speed, managing memory efficiently, and optimizing battery usage. By addressing these areas, developers can create applications that perform well under various conditions and deliver a seamless experience to users.

Improving app speed and responsiveness is the first step in performance optimization. This involves minimizing the time it takes for an app to load and respond to user interactions. One of the primary strategies for achieving this is code optimization. Writing clean, efficient code and avoiding unnecessary computations can significantly reduce execution time. Developers should also leverage asynchronous programming techniques to prevent blocking the main thread. In JavaScript, for instance, async/await or promises allow tasks to run concurrently, improving responsiveness. Similarly, AsyncTask or Kotlin coroutines can handle background tasks in Android development without freezing the user interface.

Another critical aspect of app speed is optimizing network operations. Many applications rely on network requests to fetch data, which can be a significant source of latency. Reducing the number of network calls, using efficient data formats like JSON instead of XML, and implementing caching strategies can help minimize delays. Caching involves storing frequently accessed data locally so that subsequent requests can be served quickly without fetching data from the server. Additionally, using content delivery networks (CDNs) can speed up the delivery of static assets like images and scripts by serving them from servers closer to the user's location.

Efficient memory management is essential for preventing memory leaks and ensuring an application runs smoothly, especially on devices with limited resources. Memory leaks occur when an app retains references to no longer needed objects, preventing the garbage collector from reclaiming memory. Over time, this can lead to increased memory usage and eventually cause the app to crash. Developers can use profiling tools like Android Studio's Memory Profiler or Xcode's Instruments to monitor memory usage and identify leaks. Avoiding static references to large objects, nullifying references when no longer needed, and using weak references when appropriate can help mitigate memory leaks.

In addition to managing memory usage, developers should optimize the use of system resources. For example, reusing views and objects instead of creating new instances repeatedly can reduce the memory footprint. On Android, using the ViewHolder pattern in RecyclerView adapters can improve scrolling performance by minimizing the number of view lookups and object creations. Similarly, in iOS development, developers can reuse identifiers for UITableViewCells to achieve the same effect.

Battery usage is another critical factor in performance optimization, particularly for mobile applications. High battery consumption can lead to a poor user experience and negative reviews. To optimize battery usage, developers should minimize the app's background activity and avoid unnecessary wake locks that prevent the device from entering low-power states. Efficient handling of location services is also essential, as continuous GPS usage can drain the battery quickly. Developers can use geofencing or lower accuracy settings to reduce the impact on battery life.

Another strategy for optimizing battery usage is to use push notifications instead of polling. Polling involves frequently checking for updates from a server, which can keep the device awake and consume battery power. Push notifications allow the server to notify the app of updates, reducing the need for constant background activity. Implementing efficient algorithms and data structures can also improve performance and reduce battery consumption. For example, using a more efficient sorting algorithm or choosing the proper data structure for a particular task can significantly reduce the number of computations and memory accesses required.

Graphics and animations are familiar sources of performance bottlenecks, especially in visually intensive applications. To optimize graphics rendering, developers should minimize the number of draw calls and avoid overdraw, which occurs when a pixel is drawn multiple times in a single frame. Tools like GPU Profiler in Android Studio and Xcode's Metal System Trace can help identify and address graphics performance issues. Hardware acceleration and offloading complex computations to the GPU can also improve rendering performance.

It's essential to ensure animations run smoothly without dropping frames. Animations should be kept simple and efficient, using vector graphics instead of raster images, which are more resource-intensive. Developers can use animation libraries like Lottie, which allows lightweight animations created in Adobe After Effects and exported as JSON files.

Loading significant assets, such as images and videos, can slow down an application if not handled properly. Developers should use lazy loading techniques to load assets only when needed, reducing initial load times and memory usage. Image loading libraries like Glide for Android or SDWebImage for iOS can manage image loading and caching efficiently, providing options for resizing and compressing images to reduce their memory footprint.

Profiling and monitoring tools play a crucial role in performance optimization. These tools provide insights into how an application uses resources, allowing developers to identify bottlenecks and optimize accordingly. Continuous monitoring and performance testing should be part of the development process to ensure that any changes or new features do not negatively impact performance.

Optimizing performance also involves being mindful of the user experience. For instance, providing feedback during extended operations, such as loading spinners or progress bars, can make the app feel more responsive. Developers should also consider the user's context, such as network conditions and device capabilities, and adapt the app's behavior accordingly. For example, reducing the frequency of background updates on a slow network or providing lower-resolution images on devices with limited processing power can improve performance without compromising the user experience.

In conclusion, performance optimization is a multifaceted process that involves improving app speed, managing memory efficiently, and optimizing battery usage. By writing clean and efficient code, leveraging asynchronous programming, optimizing network operations, and employing effective memory management techniques, developers can enhance the speed and responsiveness of their applications. Additionally, optimizing graphics and animations, managing significant assets efficiently, and continuously profiling and monitoring the application can help maintain high performance. Balancing these technical optimizations with a focus on user experience ensures that applications perform well and provide a seamless and satisfying experience for users.

Ensuring Security

Ensuring the security of an application is paramount in today's digital landscape, where data breaches and cyberattacks are increasingly common. Protecting user data and implementing secure authentication mechanisms are critical components of a comprehensive security strategy. These measures help safeguard sensitive information, maintain user trust, and comply with regulatory requirements. This section explores the essential aspects of protecting user data and implementing secure authentication, highlighting best practices and techniques for building secure applications.

Protecting user data begins with understanding the types of data collected and the potential risks associated with it. User data can include personal information, financial details, login credentials, and other sensitive information that, if compromised, could lead to identity theft, economic loss, or other malicious activities. To protect this data, developers must implement robust security measures at every stage of the data lifecycle, from collection and storage to transmission and disposal.

Data encryption is one of the fundamental techniques for protecting user data. Encryption transforms readable data into an unreadable format using cryptographic algorithms, ensuring only authorized parties can access it. For data at rest, such as data stored in databases or file systems, developers should use robust encryption algorithms like AES (Advanced Encryption Standard) with a critical length of at least 256 bits. Additionally, it is essential to properly manage encryption keys, storing them securely and rotating them periodically to minimize the risk of unauthorized access.

For data in transit, such as data transmitted over the internet, developers should use Transport Layer Security (TLS) to encrypt the communication channel between the client and server. TLS ensures that data transmitted over the network is protected from eavesdropping, tampering, and man-in-the-middle attacks. Implementing HTTPS (HTTP Secure) by obtaining and configuring an SSL/TLS certificate for the application's domain is crucial in securing data in transit. Developers should also enforce secure connections by configuring the server to redirect HTTP requests to HTTPS and setting the Secure flag on cookies to ensure they are only transmitted over secure channels.

Access control mechanisms are another critical aspect of protecting user data. These mechanisms ensure that only authorized users can access specific resources or perform certain actions within the application. Role-based access control (RBAC) is a common approach where users are assigned roles that define their permissions. For example, an application might have roles such as admin, editor, and viewer, each with different access levels. Implementing RBAC requires defining roles and permissions clearly and ensuring that the application enforces these rules consistently.

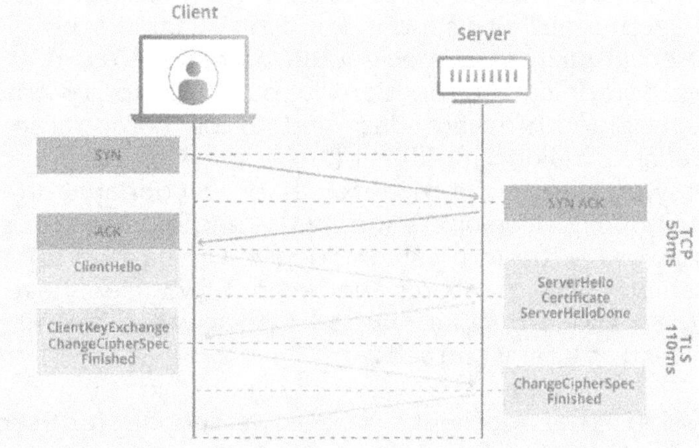

Secure authentication is the foundation of access control, verifying the identity of users before granting them access to the application. Implementing secure authentication involves several best practices, including using strong passwords, multi-factor authentication (MFA), and safe storage of credentials. Strong passwords should be long, complex, and unique, reducing the likelihood of brute force attacks and password guessing. Developers should encourage users to create strong passwords by enforcing password policies and providing feedback on password strength during registration and password change processes.

Multi-factor authentication adds a layer of security by requiring users to provide two or more verification factors, such as something they know (password), something they have (security token), or something they are (biometric data). Standard MFA methods include one-time passwords (OTPs) sent via SMS or email, authentication apps like Google Authenticator, and hardware tokens like YubiKeys. Implementing MFA significantly enhances security by making it more difficult for attackers to gain unauthorized access, even if they have obtained the user's password.

Storing user credentials securely is another crucial aspect of secure authentication. Passwords should never be stored in plain text; they should be hashed using a vital cryptographic hash function like bcrypt, script, or Argon2. These algorithms are designed to be computationally intensive, making it difficult for attackers to crack hashed passwords using brute force or dictionary attacks. Additionally, developers should use a unique salt for each password, a random value added to the password before hashing. Salting ensures that even if two users have the same password, their hashed values will differ, further protecting against attacks.

Session management is also essential for secure authentication. Once a user is authenticated, the application typically creates a session to maintain the user's logged-in state. Secure session management involves generating unique session tokens, storing them securely, and protecting them from hijacking or fixation attacks. Session tokens should be extended, random, and expire after a reasonable period of inactivity. Developers should also implement secure cookie attributes, such as HttpOnly and Secure, to protect session tokens from being accessed by client-side scripts or transmitted over insecure channels.

In addition to these best practices, developers should stay informed about the latest security threats and vulnerabilities and regularly update their applications and dependencies to address them. Using security libraries and frameworks that are actively maintained and intensely focused on security can help developers implement secure authentication and protect user data. For example, libraries like OWASP's ESAPI (Enterprise Security API) provide tools and best practices for addressing common security challenges in web applications.

Regular security testing, including vulnerability scanning, penetration testing, and code reviews, is essential for identifying and addressing security weaknesses in the application. Vulnerability scanning involves using automated tools to identify known application vulnerabilities and dependencies. Penetration testing involves simulating real-world attacks to identify security weaknesses and assess the effectiveness of existing security measures. Code reviews conducted by security experts include examining the application's source code to identify potential security issues and ensure adherence to secure coding practices.

Finally, developers should implement logging and monitoring to detect and respond to security incidents. Logging involves recording security-relevant events, such as failed login attempts, access to sensitive data, and user roles or permissions changes. Monitoring involves analyzing these logs in real-time to identify suspicious activities and trigger alerts for potential security incidents. Implementing an incident response plan ensures the development team is prepared to respond quickly and effectively to security breaches, minimizing the impact on users and the application.

In conclusion, ensuring the security of an application involves protecting user data and implementing secure authentication mechanisms. Critical strategies to safeguard user data include encrypting data at rest and in transit, enforcing access control mechanisms, and securely storing user credentials. Implementing secure authentication involves strong passwords, multi-factor authentication, and robust session management practices. Developers must stay informed about the latest security threats, regularly update their applications, and conduct thorough security testing to identify and address vulnerabilities. By following these best practices and maintaining a proactive approach to security, developers can build secure applications that protect user data, keep user trust, and comply with regulatory requirements.

CHAPTER X

Launching and Marketing Your App

Preparing for Launch

A mobile application's launch preparations include careful planning, following platform rules, and developing attractive app listings to attract users. Ensuring that programs adhere to quality standards, security protocols, and user expectations on platforms like the Apple App Store and Google Play Store largely depends on App Store guidelines and regulations. If developers want to publish their apps and attract a large audience successfully, they must comprehend and abide by these criteria.

The rules and specifications developers must abide by vary depending on the app store. These guidelines usually cover aspects including app functioning, UI design, content policies, security precautions, and regulatory obligations. For example, the Apple App Store rules strongly emphasize minimalistic design, human interface guidelines (HIG), and a clear and easy-to-use user interface. Additionally, Apple prioritizes app speed optimization, reliability, and compatibility with the newest iOS devices and versions. On the other hand, the Google Play Store's rules are concentrated on making sure that apps are safe, and secure and offer a satisfying user experience on a range of Android devices and versions.

Both app shops place a high priority on security and privacy in their guidelines. To safeguard user data, developers must put strong security measures in place. These measures should include encryption of sensitive data, safe authentication procedures, and compliance with data privacy laws like the California Consumer Privacy Act and the General Data Protection Regulation

(GDPR). Both Apple and Google mandate that apps have a privacy policy outlining the procedures for gathering, using, and sharing user data to be transparent with users regarding their data policies.

A robust app listing is essential to drawing in customers and persuading them to download the app. The app name, icon, images, description, and promotional videos are all included in the app listing. These components should be painstakingly designed to draw potential users in and emphasize the app's unique features, advantages, and value proposition. The app icon must be eye-catching and unique so that users can quickly recognize it on their devices. To give customers an idea of what to expect, promotional films and screenshots should correctly highlight the app's functionality, user experience, and interface.

To effectively convey to potential users the program's goal, salient features, and advantages, the description is essential. It should highlight how the app addresses a specific need or solves a particular problem and be brief, educational, and captivating. To make the explanation understandable and accessible to many users, developers should avoid technical jargon and utilize straightforward language. Users considering the app may find it more credible and trustworthy if it has endorsements, reviews, and accolades.

Improving the app's discoverability in the app store's search results requires optimizing the app's keywords and metadata. Developers should investigate pertinent keywords that users can use to find apps that are similar to their own. The app's title, description, and metadata should all deliberately contain these keywords to increase visibility and draw in organic visitors. Combining keyword optimization with preserving the app listing's organic readability and clarity is crucial.

Localization is another crucial component of getting an app ready for release, particularly if you want to target a worldwide user base. The app listing's attractiveness and relevance to consumers in other locales and languages can be significantly increased by localizing the app's name, description, screenshots, and promotional materials. Localizing the app's content improves the possibility that it may become famous in foreign markets and shows a dedication to catering to various user demographics.

Developers must ensure the app satisfies all platform standards and guidelines by extensively testing it before submitting it for evaluation and approval. Functionality, usability, performance, security flaws, and conformity to design specifications are tested. Before the app is made available to the general public, beta testing with a group of outside testers can offer insightful comments and helpful information about any problems or places for development. Developers should also use analytics and profiling tools to track the app's performance, find bottlenecks or inefficiencies, and optimize the program for maximum speed, responsiveness, and stability.

Developers should carefully examine and finish the app store submission procedure, which includes agreeing to the terms and conditions, uploading the necessary assets, and giving accurate information once the app is ready for submission. Software store review turnaround times can range from a few days to several weeks, contingent on the complexity of the software, the degree to which standards are followed, and the workload of the app store review staff at the moment. Developers should be ready to respond quickly to any comments or concerns from the review team and make the required changes to guarantee adherence to the rules and publishing clearance.

Following approval and release, developers should monitor the app's functionality, solicit user input, and make incremental changes based on analytics data and user evaluations. Ensuring the long-term success and longevity of the app requires maintaining continuous support and engagement with users, responding quickly to user input, and issuing updates to add new features and address bugs.

To sum up, understanding and abiding by app store guidelines and regulations, developing an appealing app listing to draw customers, and carrying out extensive testing and optimization before submission are all part of getting ready for the launch of a mobile application. By adhering to recommended methods for app store compliance, creating a captivating app listing, and verifying the app's compliance with quality and security requirements, developers can optimize the application's visibility, attract a broader user base, and attain prosperity in the fiercely competitive app market.

Marketing Strategies

A mobile application's marketing strategies influence its success significantly; this encompasses pre-launch and post-launch marketing strategies. Even before the actual creation of the app and its release, a marketing campaign is initiated to create awareness, generate interest, and cultivate the target audience. After the app release, the emphasis is placed on attracting users that will engage, become visible and download the app for steady and continuous growth.

Pre-launch marketing strategies play a pivotal role in setting the stage for a successful app launch. By delving into the target market's needs, preferences, and trends, developers can unearth key opportunities and assess the market's potential. This process, though demanding in terms of market research and competitive analysis, can pave the way for a robust and distinctive brand image.

This is crucial for creating a brand that stands out, captures customers' attention, and differentiates the app from its competitors.

Creating a pre-launch marketing strategy involves employing several channels and strategies to create awareness and discussion about the application. Through Facebook, Instagram, Twitter, and Linked In, the developers could engage with potential users to share updates, behind-the-scenes videos, and sneak peeks at creativity. Coming to specific groups and bringing traffic to the app website or landing page where they can get more info and subscribe to release alerts, paid social media advertisements, and search advertisements can be helpful.

Another successful pre-launch strategy is content marketing which involves creating and sharing valuable, informative, and engaging content on the app and its benefits and trends. Blog posts, articles, infographics, and videos can help developers define the app's value proposition, solve users' issues, and build thought leadership. Including appropriate keywords and phrases in your content optimization allows you to attract organic traffic from those interested in similar products or topics.

Email marketing campaigns can also be used before the launch of the app to build relationships and create expectations from potential customers. It is recommended that developers offer discounts or early access to their apps for users who register an account or invite friends through email newsletters, sneak peeks, and countdowns. This makes it easier for the email recipients to download the app when available since the information is personalized depending on interest and behavior. This could increase interest and sales.

As the date for the launch of the app approaches, developers should create a good listing in app stores to attract the users' attention, explain why it is helpful to download the app and encourage people to do so. This means using the title, description, screenshots, and promotional videos to amplify the software's compelling features, benefits, and unique selling points. Adding user awards, reviews, and testimonials to the page contributes to giving credibility to the app, and this may make other potential users download it. Post-launch, the focus shifts to more operational marketing strategies that drive downloads and engage the existing audience with the app.

After a launch, app store optimization or ASO comes into play. It involves constant optimization of the app listing to enhance search visibility and attract more organic traffic. This involves monitoring the keywords the application targets, analyzing customer feedback and ratings, improving graphics, and experimenting with strategies to increase downloads and conversions. Paid approaches to acquisition can improve the app's presence, reach new users, and drive installation. Some of the strategies include the use of influencers, social media marketing, and PPC marketing.

The campaigns should be selected according to the demographic, interest, and activity data to ensure maximum ROI and attract relevant users who will engage with the app in the long term. After the launch, mobile marketing mainly relies on app analytics and tracking because they provide numerous insights on user engagement, conversion rates, acquisition, and performance. They may more effectively market their products, refine their targeting methods, and constantly enhance the usability of their applications by regularly reviewing the data for trends, opportunities, and areas for progression.

User acquisition and retention are two critical elements that help in app growth and maximizing users' lifetime value (LTV). Analytics can be integrated into the app more effectively to capture the user's actions, interests, and interactions with the product. Based on the activities and interests of the users, it will be possible to send targeted messages and notifications.

Maintaining and Updating Your App

Maintaining and updating a mobile application is a continuous process beyond the initial launch phase, focusing on collecting user feedback, addressing issues, and delivering regular updates with new features and improvements. This ongoing effort is essential for ensuring the app remains competitive, relevant, and aligned with user expectations in a dynamic and evolving marketplace. Effective maintenance and updates enhance user satisfaction and contribute to app longevity, engagement, and sustainability over time.

Collecting user feedback is a foundational aspect of maintaining and improving an app. Feedback provides valuable insights into user preferences, pain points, and areas for enhancement, helping developers prioritize feature development and address issues promptly. Various channels for gathering feedback include app store reviews, ratings, customer support inquiries, surveys, and social media interactions. Developers should actively monitor these channels, analyze feedback trends, and categorize feedback to identify common themes and prioritize enhancements that align with user needs and expectations.

Responding to user feedback is equally important as collecting it. Engaging with users through timely responses to reviews, comments, and inquiries demonstrates responsiveness and commitment to addressing user concerns. Developers should acknowledge feedback, clarify misunderstandings, and provide transparent updates on addressing reported issues or implementing requested features. Effective communication fosters trust and goodwill among users, encouraging them to remain engaged with the app and contribute constructive feedback in the future.

Regular updates are critical for maintaining app performance, stability, and security and introducing new features and improvements. Developers should establish a structured release schedule based on user feedback, development cycles, and business objectives. Minor updates typically focus on bug fixes, performance optimizations, and addressing user-reported issues to ensure a smooth and reliable user experience. Major updates, on the other hand, introduce significant new features, enhancements, or design changes that add value and differentiate the app in the marketplace.

Prioritizing updates based on user feedback and data-driven insights ensures that development efforts are aligned with user expectations and business goals. Developers can use analytics tools to track app usage, user engagement metrics, and conversion rates, providing valuable data to inform update decisions. User behavior analysis helps identify popular features, underutilized functionalities, and potential pain points that can guide feature prioritization and optimization strategies.

Incorporating new features and innovations is essential for keeping the app competitive and meeting evolving user demands. Developers should continuously assess industry trends, competitor offerings, and technological advancements to identify opportunities for innovation and differentiation. Introducing unique features, integrations with popular platforms or services, and enhancements that improve usability or convenience can attract new users, retain existing ones, and enhance the app's value proposition.

Testing and quality assurance are critical to the update process to ensure that new features and improvements are thoroughly evaluated for functionality, performance, and compatibility. Developers should conduct rigorous testing, including unit testing, integration testing, and user acceptance testing (UAT), to identify and resolve issues before releasing updates to the public. Beta testing with trusted users or a dedicated test group can provide valuable feedback and insights into real-world usage scenarios, helping refine features and ensure a positive user experience upon release.

App store optimization (ASO) is another important consideration when updating an app, involving ongoing optimization of the app store listing to improve visibility and attract downloads. Developers should update app titles, descriptions, screenshots, and promotional materials to reflect new features, highlight improvements, and align with current trends and user search behavior. Optimizing keywords, monitoring competitor activity, and leveraging user reviews and ratings can further enhance ASO efforts and maximize app discoverability in app store search results.

App Store Optimization Elements

Engaging users during the update process through proactive communication and marketing efforts can build excitement, encourage adoption, and drive downloads. Developers can use email newsletters, social media announcements, and in-app notifications to inform users about upcoming updates, highlight new features, and solicit feedback or suggestions. Offering incentives, such as exclusive access to beta versions or special promotions for early adopters, can further incentivize users to update their apps and explore new functionalities.

Post-update monitoring and support are essential for evaluating the impact of updates, identifying any issues or unexpected behaviors, and addressing user feedback promptly. Developers should monitor app performance metrics, user reviews, and support inquiries following updates to assess user satisfaction, gather additional feedback, and iterate on future updates based on real-world usage and user response.

In conclusion, maintaining and updating a mobile application involves a continuous cycle of collecting and responding to user feedback, delivering regular updates with new features and improvements, and optimizing app performance and engagement over time. Developers can enhance app quality, user satisfaction, and long-term success in a competitive app marketplace by prioritizing user needs, embracing innovation, and leveraging data-driven insights. Effective maintenance and updates drive app growth and retention, foster a loyal user base, and position the app for sustained relevance and profitability in an ever-changing digital landscape.

CONCLUSION

By the time we wrap up "Mobile App Development: Crafting Innovative Digital Experiences: A Beginner's Guide to Building Your First Mobile Application," you will have gone from ideation to app store launch with a mobile application that works flawlessly. You now have the fundamental understanding, valuable abilities, and self-assurance needed to realize your creative digital experiences thanks to this book.

With the comprehensive coverage of the critical phases of app development—ideation, market research, planning, design, coding, testing, and launch—you will emerge from this book with a sense of accomplishment. You now possess the skills to tackle the challenges of UX and UI design, ensuring your application is both visually appealing and user-friendly. This book demystifies the technical aspects of setting up your development environment, writing clean, efficient code, and integrating advanced features, enabling you to create a robust and reliable program.

With a strong emphasis on security, performance optimization, and thorough testing, this book empowers you to ensure your software works flawlessly, offering a safe and satisfying user experience. As you prepare to release your app, you understand the crucial role you play in using marketing techniques and maintaining it regularly to keep it engaging and relevant.

Your journey towards developing a mobile app only begins with this guide. The potential to create novel digital experiences will increase as technology develops further. Accept the obstacles, keep learning, and allow your imagination to lead you in making valuable apps that improve and enrich people's lives worldwide.

Thank you for buying and reading/ listening to our book. If you found this book useful/ helpful please take a few minutes and leave a review on the platform where you purchased our book. Your feedback matters greatly to us.